Agricultural Extension

The Training and Visit System

Daniel Benor, James Q. Harrison, and Michael Baxter

The World Bank
Washington, D.C., U.S.A.

Library of Congress Cataloging in Publication Data

Benor, Daniel.
  Agricultural extension

   1. Agricultural extension work. I. Harrison, James Q.   II. Baxter,
Michael W. P.   III. The World Bank. IV. Title.
S544.B435   1984   630'.7'15   82-23865
ISBN 0-8213-0140-3

Contents

Foreword

Barely a decade has passed since Daniel Benor first publicized a new idea for reforming agricultural extension services. The forerunner to this edition—a booklet on *Agricultural Extension: The Training and Visit System* by Daniel Benor and James Q. Harrison (The World Bank, 1977)—became the standard of both policymakers and agricultural practitioners. This edition, revised by Daniel Benor and Michael Baxter, takes account of a decade of experience in some forty countries and presents both the lessons learned from practical experience and the accumulating evidence that the training and visit (T&V) system of agricultural extension is an excellent investment.

The central theme of the T&V system—efficiency in the use of resources available to government and farmers—has proved particularly useful in a period during which many governments had to scale down the level of new investments and recurrent expenditures. While it is difficult to measure results of the T&V system in precise quantitative terms, reforming agricultural extension has been shown to be instrumental in raising productivity or even in breaking out of stagnation. The training and visit system is not without cost to governments: Resources are needed to set up the system and operate it, and officials must overcome ingrained habits to change lines of command in building a single-purpose system of professional advice to farmers. These costs, however, pale in comparison to the benefits of fully mobilizing, through efficient management, the existing potential of prior investments by governments and farmers, both in rainfed and irrigated agriculture.

While the T&V system of agricultural extension is in the first instance concerned with cultivation practices on the farm, it reaches into other areas of governmental policy and resource allocation. The closest interaction is with agricultural research, the main source of messages recommended to, and adopted by, the farmer to increase his productivity and income. The unique method of feedback by the farmer to extension and research helps to reorient research towards solving actual production constraints on the farm. Beyond research, a well-functioning T&V system provides signals for other improvements in the farmer's production environment. The rise in farmers' incomes generates additional

v

demand for inputs and institutional credit. As farmers learn to use scarce resources optimally, it becomes easier for government to reduce price distortions and input subsidies.

The World Bank is proud to have been associated with the efforts by so many governments to introduce an effective and professional agricultural extension system. The results warrant continued support of T&V in the context of a strategy that assists in making improvements in the policy framework and specific investments for developing the agriculture sector. It is most gratifying to witness that the training and visit system of agricultural extension is not perceived as the subject of outside intervention but as a system that governments adapt and develop in a continuing exchange of experience among themselves.

Ernest Stern
Senior Vice President,
Operations

Preface

Much has happened in reformed agricultural extension since Daniel Benor and James Q. Harrison released their paper *Agricultural Extension: The Training and Visit System* (The World Bank, May 1977) seven years ago. The training and visit (T&V) system of agricultural extension, designed by Daniel Benor, has been adopted in either an explicit or implicit form by some forty developing countries in Asia, Africa, Europe, and Central and South America. Eight countries and thirteen major states in India have adopted the system in their entire area covering all farm families; other countries have adopted it in more limited areas in conjunction with agriculture and rural development projects assisted by the World Bank or by other resources.

The system emphasizes simplicity in organization, objectives, and operation. It has a well-defined organization and mode of operation, and provides continuous feedback from farmers to extension and research and continuous adjustment to farmers' needs. It has spread rapidly because of its attractiveness both as a means to increase the agricultural production and incomes of farmers, and as a flexible management tool that is well suited to the needs of departments of agriculture in many developing countries. The experience in many countries in implementing the training and visit system has suggested areas where a change in emphasis, clarification, or adjustment is required. These adjustments do not alter the basic precepts and objectives of the system, but they do take full advantage of one of the key features of an effective extension system: feedback from the field.

The very success of the system has contributed to some difficulties in implementation. While the preface to the 1977 paper cautioned readers to reflect on the reasons for the system's success before hastening to initiate similar measures, this advice has not always been heeded. In the process, some fundamental requirements for the effective introduction of the system—such as, a decisive setting of priorities, a single-minded concentration of efforts to ensure success right from the start, relevant training, and the development of appropriate technology—have often been ignored. There has also been some confusion about central aspects of the system—for example, the role of contact farmers and subject matter specialists, and the primacy of field work and farmer contact by staff at all levels—that has sometimes resulted in a less effective operation. In the light of the experience of the many departments of agriculture that have adopted the system and of the Bank's experience in working with these extension services, this is an appropriate time to revise and expand the system's guidelines.

This booklet is a revised version of the 1977 paper noted above. It includes as an Annex the chapter summaries from a more comprehensive statement on the extension system, *Training and Visit Extension* by Daniel Benor and Michael Baxter (The World Bank, 1984). Readers requiring a deeper account of the training and visit system of agricultural extension are referred to that book. This booklet may be used as a general, self-contained introduction to the training and visit system.

The revision is based mainly on experience from the system's implementation by extension services in India, Indonesia, Thailand, Kenya, and elsewhere, over the past ten years. While the strong influence of experience in India will be noted, particularly in terminology used for administrative units and staff positions, it is hoped that the appropriate local equivalents can be readily identified.

Two main lessons from the experience over the past several years in implementing the training and visit system have been particularly influential in updating the 1977 paper. One lesson is the continuing need to adapt any extension system, in this case the training and visit system, to the agricultural and administrative structure of a country. The objective of reforming extension is to establish an effective, professional agricultural extension service. For many countries, the training and visit system has proved to be such a means. For others, different systems, or adaptions of the training and visit system, may be more appropriate. A second important lesson is that, if a decision is made to adopt the training and visit system, and while acknowledging the need for adjustment to local circumstances, the basic principles of the system must be well understood and followed, and there is no room for significant variations in the system's basic features. Examples of these features are: fixed, regular visits to farmers' fields by all extension staff; the primacy of able subject matter specialists and of strong, two-way linkages between farmers, extension, and research; the development of specific, relevant production recommendations to be taught to farmers; frequent, regular training of all extension staff; and exclusivity of function (that is, all extension staff should concentrate on extension work only).

This booklet and the book *Training and Visit Extension* do not lay down definite rules on how to establish an extension system. Rather, they explain the complexity and interrelationships of training and visit extension, and draw attention to the range of considerations that are important when implementing the system. Just as experience has dictated revision of the 1977 paper on the training and visit system and indicated the need to place renewed emphasis on its salient points, so will progress with extension reform and local administrative structures and agricultural conditions suggest which parts of this booklet and of *Training and Visit Extension* require emphasis for a particular extension service.

Acknowledgements

The 1977 version of *Agricultural Extension: The Training and Visit System* by Daniel Benor and James Q. Harrison has been revised by Michael Baxter and Daniel Benor. C. M. Mathur provided valuable suggestions as to areas requiring attention, and Jacob Kampen assisted greatly by reviewing, commenting on, and discussing the revision. James Q. Harrison reviewed an early version of the revision. Margaret de Tchihatchef spent much time editing and preparing the booklet for publication.

We acknowledge and give thanks for all this assistance, without which the task would have been immeasurably more difficult.

Credit, cover photo: Directorate of Extension,
Government of India

Agricultural Extension

The Training and Visit System

ORGANIZATIONAL PATTERN OF THE
TRAINING AND VISIT SYSTEM OF AGRICULTURAL EXTENSION

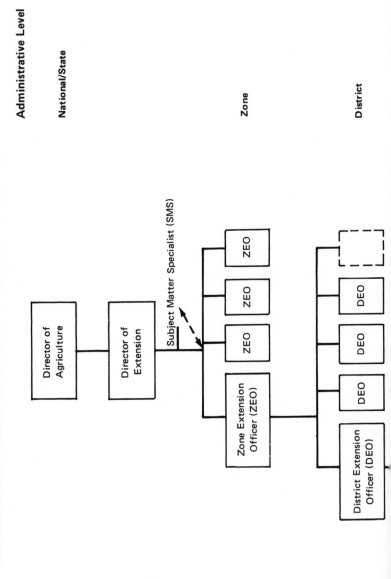

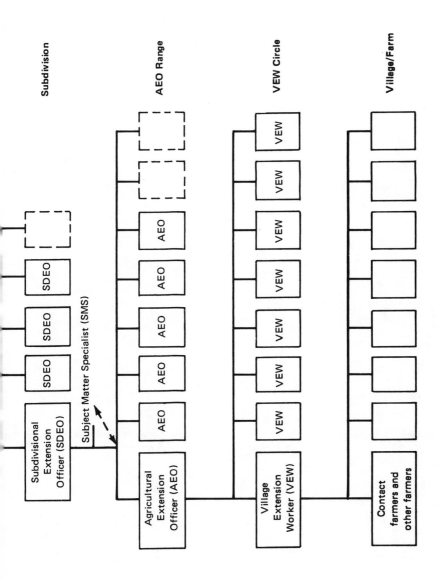

Subdivision

AEO Range

VEW Circle

Village/Farm

Subdivisional Extension Officer (SDEO)

SDEO

SDEO

SDEO

Subject Matter Specialist (SMS)

Agricultural Extension Officer (AEO)

AEO

AEO

AEO

AEO

Village Extension Worker (VEW)

VEW

VEW

VEW

VEW

VEW

VEW

VEW

Contact farmers and other farmers

3

1. Introduction

Development of agriculture is an integral part of economic development. Few countries have experienced sustained economic development without growth of the agriculture sector. Similarly, all countries that have experienced significant growth in agriculture have also achieved a more rapidly growing economy. The development of the agriculture sector is, therefore, not just an end in itself; it also has a direct and beneficial effect on overall economic development.

Government, the private sector, and farmers have key roles in bringing about agricultural development. Although the nature of their roles and the relative importance, in particular, of government's responsibilities differ between and even within countries, a main governmental role usually is to create and maintain the infrastructure required for agricultural development. Extension and research services, input supply and credit arrangements, marketing structures and price systems, as well as communication and transport networks, are basic features of this infrastructure. The private sector often has an important role in the development of such facilities, but, at the least, policy guidelines on infrastructural development and operation are the responsibility of the government.

Farmers' reactions in managing their farms and in deciding between production alternatives depend on the infrastructure and the economic incentives of the agriculture sector. In addition to establishing a supportive infrastructure, a central concern of government in agriculture development—and, by extrapolation, in economic development—is, therefore, to ensure that farmers are continually exposed to attractive production options. Farmers, small-scale as much as large-scale, react positively and quickly to attractive prices for their products. However, they cannot respond appropriately and quickly unless they clearly understand both the most recent technologies applicable to their farm as well as the broader agroeconomic environment in which they operate. To facilitate such understanding, most farmers need continuously updated advice. In view of the resource constraints faced by most farmers in developing countries, attractive production options are largely those that enable them to use their land, labor, and capital in a

5

better way. Agricultural research and extension services have a central role in facilitating this through the development of appropriate production recommendations and the transfer of new technology to farmers.

The importance of research in agricultural development and in resolving problems faced by farmers is rarely disputed: in contrast, although agricultural extension's role is as basic as that of research, the need for agricultural extension is not clearly recognized. Without local research support, agriculture will remain traditional with low yields and low productivity. Improved agriculture depends heavily on the input of research. When crop varieties with higher yield potentials have been developed, cultural practices, crop management, pests and diseases inevitably require further research support; in addition, because of ever-changing economic and physical conditions, farmers require continued research assistance. The research from which improved agriculture derives must be sound in basic as well as applied aspects; to be effective in the long run, it should be relevant to the actual production conditions and needs of farmers.

Village Extension Worker discusses recommendations with farmers on the way to the village

6

Extension also is a necessary prerequisite to widespread and sustained agricultural development. It is not possible, even in highly developed countries, to encourage farmers rapidly to adopt new technology and more efficient practices based on continuously advancing research without farmers' clearly understanding them. To bring research results and new agricultural techniques to farmers, someone must teach farmers how these practices should be employed and adopted under their own individual farming and resource conditions. Thus, an extension service is needed to explain new technology to farmers and teach them how to adapt and adopt improved production practices in order to increase their production and income. Extension also has a vital role in ensuring that the agroeconomic and social environment of farmers and the day-to-day production problems they face are appreciated by research. This feedback function of extension facilitates the continuous reorientation of research towards the priority needs of farmers and the early resolution of important technological constraints. Extension's role in agricultural development is largely catalytic and, therefore, often difficult to qualify. It is extension, however, that helps farmers take advantage of research findings and technological advances, quickly adjust to seasonal and economic conditions, and effectively use support services to increase their production and income. Without extension's guidance, farmers often are unable fully to exploit the opportunities available to them.

In many countries, extension is completely disregarded by those concerned with agricultural development. Elsewhere, it is neglected in the sense that even though an extension service is established, the standards of organization, productivity, and expectations accepted for it are below those required of many other governmental activities. If extension is as important a prerequisite for agricultural development as research, why is it often neglected by government and others?

There are two main reasons for this neglect. On the one hand, poor performance of the agriculture sector cannot easily be linked directly with a lack of extension. On the other hand, where extension works well, it is often difficult to isolate its independent impact on production. Accurate monitoring and evaluation of extension activities and its impact is methodologically difficult; in the absence of clear evidence of a direct contribution attributable to agricultural production, extension's role is frequently considered as small and unimportant.

A second consideration is that extension is one of the most difficult activities for a government to organize and administer. Farmers have a multitude of needs for external support, particularly in the early stages of economic development, and government must provide or contribute to many of these. A common response to these needs is to establish a multifunction field service, with agricultural extension as one responsi-

7

bility among many. In such circumstances, it is difficult to isolate a specific role for extension, let alone to determine its effect. Moreover, extension workers are of necessity based in scattered locations which makes it difficult to exercise regular in-field supervision and guidance to help make their work more effective. Often they are neither able nor expected to visit farmers regularly, and are made responsible for a multitude of tasks that are difficult to plan or monitor. Another reason that detracts from effective agricultural extension is the urban bias common to many governments. Together with the problems of attributing impact to extension activities, these constraints result in a situation where extension services are often neglected, the apparent assumption being that extension is something that is either not needed or which somehow will develop of its own accord.

While acknowledging the pivotal role of extension in bringing about agricultural development, it is not to say that only one form of extension organization can fulfill the required functions. An extension structure can and should be developed to suit local conditions, resources, and requirements. Thus, there are many ways in which an agricultural

Village Extension Worker cycling to meet farmers

Directorate of Extension, Government of India

extension service might be organized to serve farmers well. Even within a country, and certainly over time, there may be alternative suitable methods of organizing an extension service. However, to be effective, any system selected will need to overcome the organizational and administrative problems inherent in an extension service, while also providing an efficient tool to strengthen and operate a viable system of technology transfer and to generate meaningful feedback to research.

A professional system of extension based on frequently updated training of extension workers and regular field visits (and, therefore, called the training and visit—T&V—system) is one approach to agricultural extension that has proved to be particularly effective. The contribution of this system to agricultural extension and agricultural development is not that it is a new extension methodology (which it is not), but rather that it is an effective management system that enables the efficient implementation of known extension principles. It provides an organizational structure and detailed mode of operation that ensures that extension agents visit farmers regularly and transmit messages relevant to production needs; problems faced by farmers are quickly fed back to specialists and research for solution or further investigation; and extension staff receive the regular training required continuously to upgrade their professional ability so they may serve the technological demands of farmers.

Because the extension management system based on regular training and visits makes possible the implementation of accepted principles of effective agricultural extension, it has been widely adopted in many countries. There is, however, a danger in too rapid a diffusion of the system: some of the fundamental features of the system may then be omitted, either deliberately to achieve shortcuts, or (as is more often the case) as a result of misunderstanding of its basic principles.

A basic principle of effective, professional extension and so of the training and visit system is that farmers should be visited regularly by able and qualified extension workers. Agricultural extension is not a one-shot effort. It entails a continuous, long-term process of contact with farmers to understand their production conditions and to guide research to help develop recommendations that respond to farmers' technological needs. To do this, field visits should be regular and frequent. The best way to achieve this in developing countries is to have extension workers make field visits according to a fixed schedule that is known to all farmers. To keep extension staff up-to-date on the latest know-how and with regard to specific recommendations suited to changing farm conditions, there should be regular frequent training of staff of all levels. Linkages with research must be two-way and close, and based on joint extension/research field activities and workshops rather than on irregular meetings of committees. To establish a professional extension service, all extension staff should work exclusively on extension. Every staff member should have a clear definition of his responsibility and authority. At all levels, extension work must be field oriented: very few written reports are required.

The work of all extension staff should be focused on supporting the field-level agent, since he is the only extension worker who is in regular direct contact with farmers, teaches them extension's messages, and handles much of the feedback on their problems and reactions to the extension/research system. All staff support and assist the field worker by way of training and in-field guidance to do these tasks well. Since the business of extension is agricultural know-how, a basic characteristic of extension based on the training and visit system is the priority given to professionalism, specialist staff support, training, and close linkages with research, other sources of know-how, and agricultural universities.

There is a continuous need for orientation training of all extension staff and those of other departments involved in related development activities. Orientation is required to highlight and explain the change in the organization and operation of the extension service entailed in its establishment or reform along training and visit lines. Such change often includes the creation of a single line of technical and administrative command, the delineation of definite responsibilities and authority for each level of staff, the initiation of close in-field supervision of staff, and the elimination of most written reports, each of which is often quite different from existing practice.

While particular management principles and an organizational structure developed to meet these are central to professional extension, their effectiveness is limited without able and committed leadership. No organization can function effectively without strong leadership, and given the organizational and operational problems confronting extension, strong leadership of a professional extension service is especially vital. Management must not only fully understand the principles, operational features, and potentials of a professional agricultural extension system, it must also be able to develop and utilize efficiently the human resources available within the service, and to convey to staff and outsiders the system's objectives, functions, and achievements. High-quality leadership, and the responsibility and initiative this requires, must be developed at all levels. However, since a key to effective extension is good judgment, initiative, and adaptation to circumstances, there is no simple clear-cut pattern of leadership. Consequently, managers at all levels should be encouraged to develop their abilities of leadership by providing them with the right tools—in particular, clear management principles, operational guidelines, delineation of authority and responsibility.

In addition to the fundamentals of management principles and leadership, four points should be kept in view in establishing or reforming extension along training and visit lines. First, professional extension based on regular training and visits is not only able to serve situations of low-level agricultural development. It can be adapted to suit all levels of agricultural sophistication. The basic management principles of a professional agricultural extension service are similar, no matter what the level of agricultural technology is. Extension operations may be adjusted to meet local needs—for example, by increasing the number and level of technical specialists and field extension agents (including perhaps specialized agents for farmers who have already attained very high levels of technology), or by emphasizing the complementary support of field extension work that can be provided by well-coordinated mass media activities. The extension system should also be expanded over time to cover most farm-based production activities, although it is likely to concentrate initially on major crops.

Second, while the management principles of the training and visit system and even its basic organizational features may appear simple

Using simple teaching aids in fortnightly training

Directorate of Extension, Government of India

and straightforward, they comprise a complex system. All components need to operate well if the system as a whole is to have the desired effect. Not only must field visits be made as scheduled, but extension workers must have useful technology to teach farmers, be able to diagnose farmers' production conditions and constraints, and be effectively supervised in the field. Training sessions must teach staff how to adapt production recommendations to reflect the particular resource situations of individual farmers, build up staff capabilities so staff become able to advise farmers on the full range of their interconnected farming operations, and encourage professional interaction between extension and research.

Third, and related to the complexity of the system, is the fact that training and visit extension is a flexible system within a rigid framework. The ability to handle many different crops and other farm-based production activities as well as pest and disease problems, to adapt to varying settlement patterns and input supply and seasonal conditions, to reflect the levels of sophistication of farmers and extension staff, and to take advantage of complementary extension activities, are evidence of flexibility. Rigidity is present in the regular visit schedules and training programs, and the insistence that extension staff and supporting research services focus on relevant and economically viable ways to increase farmers' agricultural production and income. Given the range of geographic, technological, and administrative conditions under which extension activities are usually implemented, such a solid framework is necessary to ensure extension's impact.

Fourth, the introduction of professional extension based on regular training and visits entails considerable administrative reorganization (and inherent temporary disruption). In addition, professional extension often also involves radical changes in the behavior and work methods of staff, and in various key relationships—for example, between farmers and extension, extension and research, extension and agricultural support services, between the extension service and the Department of Agriculture, and farmers and "government" in general. The adoption of a professional extension system, therefore, is not a single, static decision. The system evolves as it is implemented, and implications of the introduction of the system will become apparent only gradually. The importance of continually upgrading extension staff's ability, research support, and operations of other agricultural support services, not to mention the way in which the extension service itself operates to serve farmers effectively, will be envisaged more clearly over time. Indeed, the long-term objective of extension reform may, in addition to establishing an effective extension service, include setting in motion the process of professional upgrading of the entire Department
12   of Agriculture.

Researchers and Subject Matter Specialists in a monthly workshop

B. Z. Mauthner

Professional extension cannot in isolation solve all agricultural development problems. Among other things, the whole range of agricultural support services, from the provision of improved seed, fertilizer, credit, and other inputs, to transport, communications, and marketing, must also be improved to achieve a real, sustainable impact on agricultural production. However, it is not possible, managerially or financially, to improve all these areas simultaneously. What is important is to establish an appropriate sequence in which improvement in these various areas should be tackled. A start must be made in upgrading a key service area that will generate strong demands for change in other areas and help break the vicious circle of ineffective management and inappropriate allocation of resources. This key service in many instances is extension, since an effective extension system enables farmers to use available resources in a better way, generates a higher income for them, and encourages them to demand improved services in other areas (including research). It is, therefore, logical to put priority on strength-   13 ening extension rather than attempting to improve gradually and simultaneously the entire range of agricultural support activities. A further advantage of starting with extension is that the principles on which professional extension is based are generally applicable to related management-implied activities (such as agricultural support services). Its introduction provides experience for reform of other critical activities, which, in time, will come about partly in response to farmers' pressure (which itself is one among many aspects of the impact of effective extension).

It is often argued that the introduction of professional extension is too radical and too costly a step. Moreover, the absence of a significant backlog of technical knowledge and a preponderance of staff who are poorly qualified and inadequately motivated, and have only limited mobility may make success difficult to imagine. In most circumstances, it is indeed a radical step when compared to existing extension and management systems. However, the situation that needs to be improved frequently requires profound changes. With regard to cost, a basic point is that effective professional extension involves the efficient use of available resources—both by government and farmers. If available staff, training, and other resources of government are carefully organized to provide better in-field extension support, incremental requirements may be minimized. Where suitable staff are not available, or where there are real problems of mobility and resource allocation, a start should be made on a smaller scale. This does not mean that a diffused system should be introduced over a larger area, but rather that the essence of professional extension should be adopted in a compact area. This will show the results and the impact of professional extension, and give solid evidence on which to base a decision whether further investment is justified.

When is the most appropriate time to introduce professional agricultural extension? The answer to this question is easy: it is never too early for the introduction of agricultural extension organized along professional lines. Ultimately, there must be a steady stream of improved production practices for extension to reach farmers, but even where formal research and proven know-how is lacking, there are always some farmers who are more productive than others. The reasons for the greater productivity of some farmers must be determined, as it will provide a basis for developing extension recommendations for other farmers. Moreover, unless extension begins operating systematically in the field, it is unlikely that agricultural research will develop in a way that is relevant to the needs of farmers. Farmers and extension and research staff alike will regard regular visits, training, and feedback as being far more rewarding, even with limited relevant technology available, than if no action were taken.

14

Although there are common features in all effective extension systems, the way in which these will need to be adapted to the specific conditions of any one country or region cannot be precisely determined except through some degree of experimentation. Not only is it important to build into the extension system a mechanism for on-going review and adjustment, there also should be opportunity, if possible, to experiment on a small scale before deciding how it can best be established on a more widespread basis. Such a period of experimentation is also useful in that it will allow management to appreciate some of the implicit broader organizational prerequisites and ramifications of professional agricultural extension. However, in pilot projects nothing should be incorporated into the extension system that is not successfully replicable (for cost or other reasons) on a large scale. Initially, extension reform does not need to be implemented across broad areas, but at all levels of implementation it must be economically and organizationally viable.

While appropriately designed pilot projects are useful and necessary precursors to a more widespread introduction of professional extension, it should be kept in mind that many of the prerequisites of effective extension—strong links with research and input organizations, and high levels of commitment and strong management, for example— usually only become possible (and their advantages most apparent) when the system is adopted on a countrywide or regional scale. Many development projects concentrate on small areas with small numbers of farmers. These projects often result in the most able leaders being brought to these areas, and their being supported by a disproportionate amount of funds. Rather than focusing on limited project areas, there must be broad organizational improvement and institution building to enable government machinery and development agencies to make the best use of available and new resources for the benefit of most farmers in a country.

Professional extension based on regular training and visits is an effort in this direction. It has produced good results in several countries, and sincere efforts are being made by many others to strengthen extension on similar lines. In addition to assisting in the establishment of effective, professional extension services, it is hoped that the training and visit approach will contribute to the development of improved institutional arrangements in other agricultural support services. In this way, many more farmers may benefit more significantly from the considerable manpower and financial resources devoted to agricultural development.

2. General Problems with Extension

It is difficult to generalize about the state of agricultural extension across a wide range of developing countries. Some countries have virtually no national extension service. Others have a fairly widespread extension structure that is relatively well staffed. But not many developing countries have a really effective service. There are several reasons for this. The more important ones are discussed here.

Organization

The most fundamental problem in most extension services is the lack of a well-defined organization with a clear mode of operation and the absence of a single, direct line of technical support to, and administrative control of, their staff. Frequently, the field-level extension officer reports to two or even three supervisors. Since he is one of the few government functionaries operating at the local level, it is tempting to assign various nonagricultural tasks to him. Consequently, he often falls under the control of the civil administration or a broad-based rural development organization, or both. His links with the Department of Agriculture are tenuous. The inevitable result is that he spends little of his time on real agricultural extension activities and cannot be a professional extension worker. Moreover, whatever extension work he does is neither systematically planned nor adequately supervised.

Extension agents generally do not have planned schedules of work. If extension goals are set, they are often too broad and unrealistic to achieve, are too vague to check, and bear little relevance to the local situation. Not only is extension difficult to supervise because its goals are unrealistic or vague, but in many areas no importance is attached to the systematic and constructive supervision of staff. Where supervision does take place, its objective is frequently to check that staff are at their appointed posts and that progress is being made against a number of easily measurable objectives (e.g., number of demonstrations laid out or seed packets distributed) that may have little relevance to extension's goals. Supervisors are rarely given the means or encouragement to conduct supervision of field staff in farmers' fields, or to use supervision as a means to improve the ability and commitment of staff.

Dilution of Effort

The multipurpose role assigned to the field-level extension worker is closely related to the problem of inappropriate administrative organization and supervision. The field worker is often responsible not only for all aspects of rural development (including, for example, health, nutrition, and family planning), but also for regulatory work, procurement, and the collection of a wide range of statistics. This is clearly too much for anyone, and especially for the poorly paid and inadequately trained officer so characteristic of many extension services. Moreover, the work programs themselves are often ill-defined and inadequately supported; in addition, work priorities are changed frequently. The responsibilities of the extension worker are simply too broad and vague. As a result, he can perform neither his agricultural extension duties nor his other tasks effectively, and must resort to doing only the work that is most closely monitored by his supervisors (e.g., completing reports, recording statistics, and distributing agricultural inputs) and providing service only to the most influential people in his jurisdiction (and who know his supervisors). He has little time or motivation to visit farmers' fields.

Impact points are stressed in training

Coverage and Mobility

Most field-level extension agents have an excessively large jurisdiction. They may have to cover over 2,000 farm families, and in some cases over 4,000, often spread over a large area. This already difficult task is made even more so, because the extension service usually lacks vehicles to ensure adequate mobility, and housing in or close to a staff member's area of jurisdiction is generally not available. Because no time-bound, systematic program of work exists, it is impossible to achieve the close regular contact between extension worker and farmers that is essential for successful extension. Under these circumstances, extension agents have often found it necessary and convenient to concentrate only on larger farmers; they are unable to reach the majority of farmers regularly, and large farmers can often help them attain their area-specific or quantity-oriented targets as well as facilitate housing and other amenities.

Demonstrations

Much of an extension service's field activities may be concentrated on demonstrations of particular crops or agricultural practices. Although demonstrations are conducted on farmers' fields, farmers are seldom really involved since the planning of demonstrations, the necessary supply of inputs and provision of labor, and the evaluation of results is all done by extension staff. Consequently, even where the results of demonstrations are clearly superior to local practices, the improved technology is rarely adopted by farmers because they see little relationship between the conditions under which the demonstration is conducted and their own situation. Nonetheless, because demonstrations are easily planned, monitored, and implemented (in part, because they require no input from farmers other than their land), and can be used to impress high-level officers, politicians, and others with what extension has "done," they often become an area of concentration. Unfortunately, such demonstrations are rarely evaluated in terms of the extent to which they induce farmers to adopt the demonstrated skill: it appears that their impact in this respect is often limited.

Training

Training of extension staff is usually inadequate in terms of its frequency, timeliness, and relevance. Training efforts are generally concentrated on preservice training. This, however, is often theoretical and classroom oriented, and frequently seeks to cover the whole range of crops and practices that few extension workers can be expected to remember for long. Where extension agents are multipurpose workers, their training is further diluted by nonagricultural subjects.

Once extension workers have started work in the service, little effort is made to update their knowledge of new technology or developments in agriculture, even though research information continues to be generated and farmers and agriculture must become increasingly specialized and sophisticated. Nor is much work done to upgrade the quality of extension staff in a broader sense through systematic refresher training or, for some, through university degree studies. Even where refresher training is given, important fields like extension methods, communication methods, farm management, and management skills are rarely covered.

Monthly workshop participants check crop condition in the field

Directorate of Extension, Government of India

Unspecialized Staff

Extension staff at all levels must perform a variety of nonagricultural tasks that dilute the attention they can give extension. As extension is only one of the numerous functions of an Agriculture Department, extension staff may need to move completely out of extension (e.g., into input supply, marketing, or soil conservation) in order to be promoted. For similar reasons, staff may enter middle- or senior-level extension positions from other services with neither experience nor interest in extension. Even specialized positions such as Subject Matter Specialists may be occupied by officers without appropriate qualifications, ability, or interest.

Lack of Ties with Research

Usually no effective link exists between extension and agricultural research, a situation that is detrimental to the effectiveness of both. Research and extension staff do not conduct joint field experiments, although these can readily assist either to understand the concerns and constraints of the other, as well as significant problems faced by farmers. Without a close link with extension and feedback from the field, research becomes excessively academic and does not relate to farmers' real problems. This leads researchers to focus on technically optimal situations rather than on the economies encountered under practical field conditions. Consequently, the recommendations of the extension service are often inappropriate to the farmers' needs and their technical and financial capabilities. Indeed, without a continuous flow of new, practical recommendations suited to those needs, the extension service rapidly runs out of anything to extend.

Status of Extension Personnel

All these factors result in an extension service having a low status, low morale, and low pay. Farmers then have little respect for their extension agents, whom they rarely see anyway. After many years without success or recognition, extension agents have lost much of the enthusiasm and commitment they once may have had. A vicious circle develops in which lack of success undermines the extension agents' self-confidence, making it even less likely that they will have a significant impact. Their meager salary reflects their low status and productivity. Since they contribute little, it is difficult to make a case for raising their salary, developing realistic promotion opportunities, or improving their employment conditions in other ways. Indeed, it is difficult to see how such low-status workers, ineffective at the field level, could fit into a professional cadre of extension workers that must reach to the highest levels of the Agriculture Department.

Duplication of Services

Given these shortcomings of extension, it is not surprising that many succumb to the temptation to try to increase agricultural production through special efforts outside the extension service (although this, of course, only serves to weaken extension further). One such approach has been to develop special schemes focusing on particular crops, areas, or techniques. Financial and staff resources are diverted from the regular extension service for such schemes, which often duplicate the work regular staff are supposed to be doing, and lead to confusion and resentment in the extension service. The creation of many special schemes can become quite expensive. Given financial constraints, this results in their being limited in resources, which usually means that they end up serving only a few favored farmers rather than the bulk of the farming community. Such schemes not only dissipate efforts, but also are not mutually reinforcing and obscure the need for basic reform of the extension service.

This catalog of commonly encountered weaknesses strongly suggests that the pace of agricultural development could be greatly accelerated if a fundamental revitalization of extension efforts were to take place. Attempts to cope with these weaknesses in a piecemeal fashion have, in most cases, met with little success and often have made the situation worse. What is required is a comprehensive approach that tackles simultaneously all key constraints to effective agricultural extension. The principles on which such a reform must be based are described in the following chapters.

3. Reforming Extension: Basic Guidelines

For an agricultural extension service to make a start, it must have advice to offer farmers. Almost always there is a gap between what farmers potentially can achieve in their fields and what they actually do achieve. Once the existence of such a gap is established, it is the task of the extension service first to close the gap and then to be the prime initiator for generating additional know-how. Such gaps are usually apparent in all areas and crops. Where research findings are not available to help bridge the gap, extension can work on eliminating the difference between what a few good farmers do and what the rest of the farmers in the area practice. In the rare cases where gaps do not exist, extension should guide research to focus on constraints felt by farmers by bringing farmers' problems to research, and, in cooperation with research, conducting field experiments. Extension has a role wherever a difference exists between what is done and what can be done on farmers' fields: even where the difference is slight or nonexistent, extension has a role of pushing research. The extension service should be organized to deal with a wide variety of situations effectively. This applies to irrigated and rainfed agriculture and mixed farming with livestock, as well as all other aspects of agricultural production in farmers' fields.

An extension service may be organized in a variety of ways to serve farmers efficiently. Particular administrative traditions and agricultural and social conditions may require certain forms of organization. To be effective, however, there are a number of distinct features that successful agricultural extension systems must share. Basic guidelines for the reform of extension along training and visit lines are outlined here; detailed features of the training and visit system, which are derived from these guidelines, follow in the next chapter.

Unified Extension Service

One of the most essential management principles to be adopted in creating an effective, professional extension service is the establishment of a single line of command from the apex of the government agency responsible for agriculture to the field-level extension worker. Unless this agency (in many cases, the Ministry or Department of Agriculture) has full administrative control of the extension service, it is not possible to carry out extension systematically and effectively because the concerns of other agencies involved will continuously interrupt the program of extension work. However, control of field-level extension staff often does not lie with the Agriculture Department. Therefore, one of the first and most difficult tasks that faces those seeking to revitalize an extension service is the transfer to the Department of Agriculture of full administrative control of field-level extension agents and all other agricultural extension staff. This transfer is often easier to achieve when it is pointed out that extension agents responsible to more than one department achieve little either in agriculture or in their other developmental or administrative activities, and so the agency from which they are transferred has little to lose.

Although the political and administrative difficulties involved in such a transfer are substantial, they can be overcome, and the results frequently justify the transfer. Once the single line of command is introduced in part of a state or country, its success invariably generates pressure for wider application, making it easier to achieve the transfer of all necessary staff. Not only do extension managers like the system because they have full control over their staff, but extension workers at all levels prefer it because having only one department to serve, they are able to concentrate exclusively on extension and, thus, achieve greater job and professional satisfaction than previously possible.

All agricultural extension activities should be combined into a unified extension service. Staff engaged in special crop or area development schemes should be merged with the regular extension service staff. Notwithstanding the fact that the initial focus may be limited, the ultimate goal of reorganization is to develop a single, modern, professional service capable of providing farmers with sound technical advice on their entire farming operation. If the service does this, special schemes are not needed; if it does not, extension should be strengthened and improved until it can. Specialized technical advisory services may be justified in particular situations, as when an area's agriculture is dominated by a crop that is the raw material for a vertically integrated industry (e.g., as is often the case with rubber, tobacco, or tea). However, where such crops form part of more diversified cropping patterns, specialists in these crops should work through the unified extension service. Although there should be only one extension service, this does not necessarily mean that each extension agent will cover all crops or farm-based production activities. For example, important commercial crops or livestock may ultimately be covered by specialized agents—and will certainly have specialized technical support for field

The extension worker uses a variety of means to advertise his visit schedule and activities

B. Z. Mauthner

extension workers—while other agents cover a variety of different crops. However, all extension agents, no matter how specialized, should operate within the one extension service in order to provide coordination and the requisite synthesis between production recommendations taught to farmers and to ensure that the service operates effectively and efficiently.

Extension Exclusively

Extension personnel should devote all their time exclusively to professional agricultural extension work. They should not be assigned responsibility for regulatory functions, supply of inputs, and collection of statistics, and the like—not to mention other assignments that have no connection with agriculture. Such activities, which often have to be performed in the peak agricultural season when extension staff are most needed by farmers in their fields, will consume much of extension staff's time and divert their attention from their main responsibilities, undermine farmers' trust in them, and interfere with their necessary systematic and time-bound plan of work. Moreover, extension staff are not trained to perform these functions, which should be the responsibility of specialized staff, just as agricultural extension is the sole responsibility of professional extension workers.

Systematic Training and Visits

Once a single line of command has been established for a unified extension service and extension personnel are able to devote all of their time to extension, the work of the service needs to be organized in a systematic time-bound program of training and visits. Schedules of work responsibilities and training must be clearly specified and closely supervised at all levels. All staff must have realistic work loads and access to frequent, relevant training.

The field-level extension worker—the Village Extension Worker (VEW), who like all extension workers may be either a man or women (though, in many countries, most are presently men)—is responsible for a manageable number of farm families who are then divided into eight "farmers' groups." The VEW meets each of these groups on a fixed day once each fortnight, and he himself is trained on one day each fortnight by Subject Matter Specialists (SMSs). The immediate supervisor of the VEW—the Agricultural Extension Officer (AEO)—is responsible only for a manageable number of VEWs (about eight). He meets the VEWs regularly in the field each fortnight, where he ensures that they are making their scheduled regular visits to farmers, doing proper extension work, and are being well trained. The AEO is also regularly trained each two weeks (with his VEWs). Subdivisional Extension Officers (SDEOs), the supervisors of AEOs, have a manageable number of AEOs to supervise and regular training responsibilities. Similarly, Subject Matter Specialists (SMSs) receive regular training by research and have a fixed, regular work program.

Extension agents receive regular and intensive training in the most important production recommendations for the coming fortnight of the crop season. With this concentrated training, which is received in manageable, fortnightly doses, extension agents develop a thorough understanding of what they must teach farmers at a given time; this increases their self-confidence and makes them better able to persuade farmers to adopt recommended practices.

25

Concentration of Effort

The whole approach to establishing an extension system should be permeated by a concentration of effort to achieve a clearly visible impact at an early stage and thereafter continued progress. Such concentration of effort is a feature of all aspects of the system. Extension staff work only on agricultural extension and do not spread their efforts over a wide range of other activities. Within agriculture, they concentrate initially on the most important crops and, for those crops, on a few practices that bring the most immediate economic results. Extension workers concentrate mainly on a few selected farmers, not to favor the few but rather to produce the impact that is required to spread practices to the majority of farmers most quickly. Training sessions will emphasize the most important points to be handled in the next few weeks. Efforts are concentrated on making the training of extension staff, and their visits to farmers, effective in the shortest possible time. Thus, concentration is always on a few most critical points; as results are attained, the focus will change.

At the initial stage of extension reform, it is very important to make a rapid impact that will give farmers confidence in the extension staff and the staff confidence in themselves. Once this is attained, the process is self-reinforcing: farmers will expect and demand more and more from extension, and extension staff will be motivated to work harder to achieve further success.

Imitable Contact Farmers

It is impossible to maintain regular personal contact with all farmers. In good extension work, however, this is neither necessary nor desirable. The messages of the extension service should be focused mainly on selected contact farmers and other interested farmers: their fields where practices recommended by the extension agent are adopted will speak for themselves and encourage other farmers to try these practices. Consequently, three main criteria in selecting contact farmers are that they must be willing to try out practices recommended by extension, be prepared to have other farmers visit their fields to observe their practices, and be accepted by other farmers as reliable sources of advice on farming.

Technical advice spreads from the extension agent through contact farmers and some other farmers who follow extension's advice to large numbers of farmers mainly by two mechanisms. First, other farmers see what contact farmers try in their fields and the results they achieve. This generates interest. Second, each contact farmer talks about the practices he has been taught to several friends, relatives, or neighbors, and thereby helps them understand and adopt the recommendations. In this way, a large proportion of farmers can be quickly reached. Experience in areas where this system is operating indicates that within two years or so, over half the farmers are aware of, and following, some of the newly recommended practices.

Contact farmers must be selected with great care given their critical role in extension: it is best that they be selected by the extension workers in consultation with village leaders. They should not be only the community's most progressive farmers, since neighbors usually regard these farmers as exceptional and tend not to attempt to imitate what they do. On the other hand, very weak farmers tend to be slow in adopting new methods. Contact farmers must be of good standing in their community so that their views on new practices will be respected by other farmers. All major agricultural, economic, and social conditions in a farmers' group should be represented by one or more contact farmers, as each farmer in the group must have at least one contact farmer whom he regards as imitable. If it turns out that a contact farmer is not interested in the work of extension, or is in other ways unsuited, he should be changed.

All farmers should be aware of the fixed day when their area will be visited by the extension agent and should be encouraged to attend the visit. On a visit, however, the VEW focuses on contact farmers and other farmers who are interested in seeing him, because it is not possible (nor necessary) to give to every farmer individually the attention required to teach and encourage the adoption of recommended practices. The focus on contact and other interested farmers is not to help only these farmers, but to use them to convince all farmers in the group of what everyone can achieve. Any farmer can be present—and should be encouraged to be present—when the extension worker visits a contact farmer; by doing so, he will automatically learn everything the contact farmer is taught. In addition, the extension worker will contact directly some of the other farmers on every visit because, for example, they may have shown particular interest in his work, have a problem on which they need advice, are growing a crop that contact farmers are not, or because he just happens to walk past them in their field.

Best Use of Available Resources

A fundamental concept underlying professional extension is to teach farmers to make the best use of the resources available to them. This concept influences the content of the message that the extension agent teaches the farmer. Exactly what this message should involve needs to

Summaries are handed out in fortnightly training session

Directorate of Extension,
Government of India

be determined and checked thoroughly in the field. Nothing should be recommended that will not significantly and visibly increase farmers' incomes and efficiency, make their work easier or less costly, or produce better crops. In most areas, the initial concentration of the reformed extension service should be on improvement in low-cost agricultural management practices. These include better land preparation, improved seedbed and nursery maintenance, use of good seed and appropriate improved varieties, seed treatment, timely operations, weeding, proper plant spacing, and so on.

Initial emphasis should be on better management practices for several reasons. First, these improved cultural practices are well known to produce dependable results, which means that farmers face little risk in adopting them. Second, while their adoption may require more work, they require little cash outlay. Since most small farmers have surplus labor but little cash, this fits their financial requirements. Finally, farmers often cannot get the full (or even the minimum required) benefits from purchased inputs or more costly investments (such as tubewells) until and unless their basic practices have improved. For example, if farmers apply fertilizer to a field that is not properly weeded, in effect they fertilize weeds, which can then compete more successfully for available soil moisture, nutrients, and light: the result can be a lower, rather than the intended higher, yield.

Farmers can usually increase yields and incomes substantially by initially concentrating on low-cost management practices. This gives them confidence in the extension agent and makes them more receptive to his advice. Since these practices are normally well known and tested, they can be fed into the extension service quickly without requiring elaborate, time-consuming screening and trials. With such recommendations, in many agricultural areas the extension service can almost immediately obtain the initial impact that is fundamental to become firmly established and have continued success.

Even if the recommended practices involve little or no monetary investment, farmers should normally be encouraged to adopt them at first on only a small part of their land. This reduces risk and hesitation, and allows the farmer and others to compare the results of the improved practice with traditional practices in the same field. If the recommended practice is successful, farmers will rapidly extend the area under it the next season, even without waiting for further extension advice, and extension staff can then focus on other practices. If the practice fails, little is lost and it can be referred back to research.

Another aspect of making better use of available resources is to take best advantage of the resources of the extension service itself. However, many of the staff that become available for field work, technical support, or supervision are usually not well prepared. Careful organization, clear definition of responsibilities, fixed work schedules, and frequent and regular training are required: with this there can be significant results even in the early stages of extension reform.

Recommendations According to Farmers' Ability

An initial focus on management practices has been found well suited for a wide variety of rainfed and irrigated conditions in a number of developing countries where yields are low and practices need improvement. There are areas in all countries, however, where yields and management practices have already reached fairly high levels, and this must be reflected in the recommendations (and other activities and organization) of the extension service. Recommendations must be tailored according to the ability of the farmers and, therefore, should include appropriate advice on purchased inputs. Not only between regions, but even within one extension agent's area, there may be several farmers whose practices are considerably more advanced than

Many farmers gather around the extension worker as he demonstrates in the field

30

S. L. Ghosal

those of the other farmers. While the extension agent should concentrate on the kind of advice needed by the majority of farmers, he should also spend time with more advanced farmers, for they will provide an example of what most farmers will probably be doing a few seasons later.

After improvement in his crop management practices, the average farmer's income will be higher, and he will be better able to adopt more expensive and more "radical" recommendations (e.g., fertilizers, or new crops, crop varieties, or cropping patterns). Even as farmers move to this stage, however, extension's emphasis should remain on how best to use whatever purchased inputs can be afforded, rather than on the optimal amounts of inputs. The application of chemical fertilizer provides a good example.

Extension agents frequently recommend only set quantities of N, P, and K as "optimal" for a given area. Apart from the fact that these doses are often not optimal (they are generally high) and are seldom adjusted either to the specific fertility level of an individual farmer's field or the prevailing input-output price ratios, few farmers can easily afford such amounts. A more practical approach is to recommend that farmers try on a small area the minimum quantity of fertilizer that would noticeably increase their yields and incomes, and to teach them how to make the best use of this amount—when and how to apply it, and how to combine it with the use of organic fertilizers. In subsequent seasons, the farmer could gradually increase the amount of fertilizer to a more optimal level over a larger area. All of extension's recommendations, of course, should have as their basic goal an improvement in farmers' net incomes by increasing their production or efficiency (and so reducing cost and effort).

Linkages with Research

The initial messages of the extension service should take advantage of the gaps between existing agricultural practices and the backlog of research findings that have not yet reached the farmers. Such gaps are now large, but with time effective extension can close many. Therefore, in order to remain effective, extension must be linked to a vigorous research program that is well tuned to the needs of farmers. Without results from a network of regional research stations and trials, upon which new recommendations can be based, and without continuous feedback from the field to research, extension would soon have very little to offer farmers, and research would lose touch with the real and changing problems farmers face. Promotion of the necessary effective links and interaction between extension and research is an important objective of extension reform. This is achieved by regular training sessions and workshops for extension and research personnel, joint participation in planning each season's extension and adaptive and applied research activities, shared responsibility for farm trials, joint field trips to review specific crop problems and to obtain a better idea of actual production conditions, and visits by extension staff to research stations.

Supply of Agricultural Inputs and Credit

The functions of the extension service and agricultural input and other service agencies must be clearly delineated. Extension requires an effective input supply system, and input agencies need extension to advise farmers of suitable inputs and to teach them how to use these. But just as input supply, credit, and other support services are not responsible for agricultural extension, neither can the extension service be responsible for ensuring the availability of inputs to farmers, filling in loan applications, collecting debts, and so on.

While not being responsible for the actual supply of inputs, an extension service assists input agencies in a number of ways. First, it helps generate demand for purchased inputs, which in turn increases the business volume and viability of input agencies. Even while concentrating on management practices, an extension service does recommend some purchased inputs to some farmers. As their needs increase, pressure builds up from farmers for adequate and timely input supplies. Second, extension provides farmers with information on where to purchase inputs and their prices, how to use inputs profitably, where and how to apply for credit, and how such credit and inputs can be used to augment their incomes. Third, extension staff provide supply agencies with rough estimates of the area's demand for inputs.

While extension and input agencies have different responsibilities, their mutual dependence requires effective and continuous coordination. The extension service needs to know what inputs will be available and when; input supply agencies must know the extension service's areas of priority and likely recommendations involving inputs. Coordination must take place at all levels of extension planning, as well as in training sessions. Representatives of input agencies should attend extension planning and training sessions, and be particularly notified of any sessions addressing input-related questions. Detailed discussions of likely input demand, and the current and probable supply situation, should be a feature of extension training sessions. Experience with extension services indicates that with such involvement and coordination, input supply agencies can respond quickly to the demands generated through extension.

32

Continuous Improvement

The agricultural extension service requires a built-in process for continuous adaptation to changing conditions. Thus, an ongoing process of self-evaluation is necessary to enable the extension service to identify in time the areas needing change. As new crops are introduced, government priorities are modified, cropping patterns and agricultural sophistication change, and yields increase, extension activities and organization will also require continuous modification. The degree of specialization of extension staff may be raised; frequency of visits may be changed; in-service training may need to be more specialized and comprehensive; qualifications for new candidates will need to be more exacting.

It is also important that the extension service benefits from objective monitoring and evaluation of its field activities. This may be done by an organization either within or outside the Agriculture Department, as long as the organization selected is consistently able to provide prompt and objective information on performance against key indicators (such as the regularity of field visits by extension staff; farmer awareness and adoption of recommendations; and changes in production, yields, and incomes). Statistically valid sample surveys and crop-cutting experiments will be the main sources of information on extension field activities and production, yield, and income data. Monitoring and evaluation results should be used as a regular management tool and as feedback to improve the entire system. They should be reviewed by extension management as soon as they become available so that extension activities may be adjusted as necessary.

4. The Training and Visit System: Main Features

The basic feature of the training and visit (T&V) system of agricultural extension is a systematic program of training for the Village Extension Worker (VEW), combined with frequent visits to farmers' fields. In the field, the VEW teaches farmers recommended agricultural practices, shows them how to implement these practices, motivates them to adopt some on their fields, and evaluates production constraints and advises farmers how to overcome them. The system is organized to give the Village Extension Worker every fortnight intensive training in those specific agricultural practices and recommendations that relate directly to farm operations during the coming weeks, and to provide him with suitable technical and supervisory guidance to enable him to teach these recommendations well to farmers. The VEW visits once a fortnight, on a fixed day known to all farmers and his supervisors, each of the eight small groups of farmers with which he works. Other staff, all of whom support in one way or another the work of the VEW, have similar fixed work responsibilities and training. How the system works and the responsibilities of staff at each level are explained here.

General Organizational Structure

The entire organization of the training and visit system of agricultural extension is based on the total number of effective, operating farm families, and on the number of families that one VEW can be reasonably expected to cover. Once this is determined, the number of VEWs needed for a given area is easily calculated.

All other staff requirements are broadly contingent on the number of VEWs. It is taken that an effective span of control for supervision is about eight. With this in mind, staff requirements are readily calculated. One Agricultural Extension Officer (AEO) guides, trains, and supervises six to eight VEWs. Six to eight AEOs are, in turn, guided and supervised by a Subdivisional Extension Officer (SDEO), who is supported by a small team of Subject Matter Specialists (SMSs). Four to eight SDEOs are supervised by a District Extension Officer (DEO), who

34

is also supported by a small number of SMSs. Depending on the number of districts, the DEO is either supervised directly from extension headquarters or from an intermediate level, by the Zonal Extension Officer (ZEO). At headquarters (and sometimes also at the zonal level), a small team of SMSs provides specialized technical support to extension management, consisting of the director of the extension service and his deputies. The main objective of this staff structure is to ensure that each level of the service has a span of control sufficiently narrow to afford close personal guidance to, and supervision of, the level immediately below.

To provide a concrete and realistic illustration of the organizational arrangements involved, the following paragraphs describe how these basic principles have been applied with specific reference to India. However, this illustration should not be considered a rigid framework as the general approach can be adapted readily to other situations. In introducing reformed extension along training and visit lines in a country or region, special attention should be given to the general administrative structure of the area. As far as the basic features of the training and visit system allow, the reformed extension service should fit the existing administrative organization. This will minimize any confusion and resistance that might accompany the introduction of the system, and will help make full use of the facilities already available at existing administrative levels.

Field Level

The number of farm families a VEW can cover will vary from place to place depending, among other things, on the VEW's mobility, topography, weather during the main agricultural seasons, settlement patterns, density of population, density and pattern of roads, intensity and standard of cropping, types and diversity of crops, and types of farming systems. The accessibility of farmers to the VEW is perhaps the most critical factor, but there is no simple, rigid rule for weighing all these factors to determine the "right" ratio of farm families to a VEW. This is a matter of careful judgment based on field visits and experience; also, the optimum ratio may change over time as agriculture, extension, and accessibility change.

Where the population density is high, with many small farmers living closely together, and only a few main crops dominate agriculture, a ratio of one VEW to about 800 operating farm families will generally be adequate; in some areas with very good farmer accessibility, a more suitable ratio could be one to 1,200 or more. Where the population is dispersed or farms are larger, a ratio of 1:500 may be appropriate. In some special cases, as in sparsely settled or mountainous areas, a suit-   35

Measuring proper seed placement

able ratio may be 1:300. The number of actual operating farm families is taken as the denominator of the VEW:farm family ratio as they normally represent the basic farm decisionmaking unit.

The number of farmers will generally be larger than the number of actual operating farm families since, for example, a farmer and his grown sons normally represent only one operating farm family and a joint family consisting of a few farmers (e.g., brothers) working together is considered as one operating farm family. In India, normally between 60 percent and 70 percent of "farm families" recorded in official records are actual operating farm families for the purpose of extension.

The operating farm families in a VEW's area of jurisdiction—the "circle"—are divided into eight groups of about equal size. Such factors as geography, size of the village, and ease of communication within the circle are taken into account when forming these eight farmers' groups. The VEW will live in his circle. In consultation with village leaders (but not exclusively on their advice), the VEW will select about ten farmers in each group to be contact farmers.

The VEW visits each of his eight farmers' groups according to a fixed schedule of visits that must be known to all farmers and supervisors. Several variations of these schedules have evolved. The most common one is a fortnightly program of visits, by which the VEW goes to each of his eight groups for a full day once each fortnight. As the schedule is fixed, the VEW always knows in advance when he will be with a particular group; and farmers always know when they are to be visited and are entitled to the in-field guidance of the extension worker.

On each visit, the VEW tries to see at least most of the contact farmers of that day's group and some other farmers as well. All other farmers in the group are also welcome to meet the VEW on this day; any of them may participate in his visits to fields and in discussions. The VEW visits farmers' fields, discussing, explaining, teaching, and demonstrating the recommendations for that fortnight, checking progress, and helping with any technical problems farmers may have. If the VEW knows the answer to such problems, he gives it; if not, he should bring the topic up at his next training session. The VEW should at once notify his immediate supervisor—the Agricultural Extension Officer—of any emergency, such as a virulent pest attack.

Another possible visit schedule is for the VEW to visit two farmers' groups a day. In this case, he will visit each group every week for half a day. A weekly visit schedule may be appropriate in intensively irrigated areas with diverse cropping patterns, or at the peak season of pest attack. Still another possibility is to shift from a weekly to a fortnightly schedule and back again, depending on the requirements of the cropping season.

Village Extension Worker. The intensive series of fortnightly (or weekly) visits, in accordance with a fixed schedule known to all, results in the farmers themselves acting as supervisors of the Village Extension Worker. Each week, the VEW devotes four days to visits, so he covers his entire circle of eight groups in a fortnight. He will spend one of the two remaining working days each week in in-service training. This in-service training is a crucial part of the extension system, since it is through these training sessions that the VEW learns what to teach farmers during his coming visits. The main training session each fortnight—the fortnightly training session—will be conducted by the team of Subject Matter Specialists responsible for the area; the other training will be organized by the AEO of a given VEW's circle. At the fortnightly training session, the VEW also has an opportunity and obligation to bring farmers' problems to the attention of the more experienced and specialized extension staff who are his trainers.

The fortnightly training is the most important training for VEWs. The sessions are scheduled and organized so that the VEWs and their AEOs are trained together for a full day in groups of not more than about thirty persons. The sessions cover only the crucial farming operations for the coming two weeks, but these are covered thoroughly. No more than one half of a fortnightly training session should be spent in lectures: the remainder should be devoted to practicing the skills required to implement recommendations, rehearsing the presentation of messages for the coming fortnight, and preparing simple visual aids to support the recommendations. VEWs should be given pamphlets summarizing the recommendations, as well as samples (of new seed varieties, for example) that they may use in their contacts with farmers. The goal of these training sessions is to make the VEW a "subject matter specialist" on the few points of particular relevance and importance to the farmers in his circle during the coming fortnight; at the same time, his technical ability will gradually be upgraded, so that eventually he is able to advise farmers on their total farming system.

While VEWs and AEOs are not responsible for distributing inputs or monitoring their use, it is important for them to know whether and which inputs are available in their areas. Input supply agencies must also know what inputs farmers will most probably require. The same is true for suppliers of other agricultural services, such as credit agencies. For this reason, the representatives of credit, input, and other agricultural service agencies, as well as marketing organizations, should participate in fortnightly training sessions. Such matters should normally not take up much time of a session, and should be on the program at a generally known, regular time.

The other regular training session during each fortnight is the meeting conducted by the immediate supervisor of the VEW—the AEO—for the six to eight VEWs in his charge. This should be a fairly informal session during which points raised in the previous week's fortnightly training session are reinforced and/or clarified, and problems encountered in the field by VEWs are discussed and either resolved or noted for taking up with Subject Matter Specialists at the following week's training session (unless there is an emergency situation requiring immediate advice). Modification of recommendations to make them better suited to local conditions should be discussed at these meetings. Whatever small amount of office work or reporting is necessary can be done at this time also, but this is not an administrative meeting.

This schedule of visits and training takes up ten of the twelve working days a VEW has each fortnight. During the remaining two days, the VEW makes extra visits to review farm trials, arranges special extension activities such as field days, makes up visits missed because of illness, holidays, and so on, and does any necessary reporting work. Each VEW keeps in his diary (in which he records on a daily basis his activities and field progress made, as well as problems encountered by farmers) a

table covering one fortnight showing his fixed days for visiting each group and for other activities. This timetable applies to every fortnight. The Agricultural Extension Officer keeps such tables for each of the VEWs he supervises. A typical table is shown.

Typical Fortnightly Visit and Training Timetable for a Village Extension Worker

First week							Second week						
Mon	Tue	Wed	Thu	Fri	Sat	Sun	Mon	Tue	Wed	Thu	Fri	Sat	Sun
1	2	MTG	3	4	EXT	H	5	6	TRA	7	8	EXT	H
		AEO			VIS				SMS			VIS	

Explanation:

 1-8 = Number of farmers' group being visited.

 MTG = Meeting of AEO with his VEWs.
 AEO

 EXT = Extra visits for checking farm trials, holding field days,
 VIS   making up visits missed because of holidays or illness, and so on, and for doing reporting work.

 TRA = Fortnightly training session (of AEOs and VEWs),
 SMS   conducted by SMSs.

 H = Holiday.

Agricultural Extension Officer. Each Agricultural Extension Officer supervises, and provides technical support to, the six to eight VEWs who comprise his "range." The AEO spends two days each fortnight in training sessions for VEWs (the meeting he himself conducts and the fortnightly training session given by SMSs) and eight days in the field supervising the VEWs and assisting them in teaching farmers the current recommendations. The AEO should have adequate transport to enable him to visit one or two VEWs in their farmers' groups each visit day. Visits made by the AEO take place in farmers' fields where he checks the impact of extension with the ultimate element of the extension system, the farmer himself.

The AEO's visits to farmers' groups should be scheduled for a month at a time and be generally known to the VEWs; they should be so arranged that over a period of several months he sees each VEW in a number of his groups. Over time, the AEO adjusts his schedule of visits to the strengths and weaknesses of his VEWs. AEOs (as well as their higher supervisors) keep a daily record of observations they make during their supervision visits in their diaries. Like the VEW, the AEO lives within the area of his jurisdiction.

Subdivision Level

The subdivision (or subdistrict), the administrative unit that comprises the area of operation of six or eight AEOs, is under the control of a Subdivisional Extension Officer (SDEO).[1] He supervises AEOs and VEWs and is in overall charge of the extension program within his jurisdiction. One SDEO can supervise effectively about six to eight AEOs. Subdivisions with large farming populations and, consequently, more VEWs, may have two to three times this number of AEOs. In such cases, the SDEO should be supported by one or more Assistant Subdivisional Extension Officers (ASDEOs), or the subdivision should be divided into additional subunits for extension purposes, each of which would be under an SDEO. Where there are Assistant SDEOs, each should supervise a circle of six to eight AEOs, leaving the SDEO only four to five AEOs to supervise, as the SDEO should also supervise the work of the Assistant SDEOs. Both the SDEO and ASDEO should be resident within the ranges of the AEOs they supervise.

Guided by the known monthly visit programs of their AEOs, the SDEO and ASDEO draw up monthly schedules of their supervision field visits. Their field visits should normally coincide with those of an AEO (and so also with those of a VEW); they should visit one or two AEOs a day. SDEOs and ASDEOs participate in all fortnightly training sessions taking place in their subdivision (as long as the dates do not coincide), as well as in all monthly workshops of extension and research staff.

The SDEO has a team of Subject Matter Specialists assigned to his subdivision. Each team has initially at least three specialists, one each for agronomy and plant protection and a Training Officer. The work of SMSs is divided into three equal parts: training VEWs and AEOs (primarily at fortnightly training sessions); making field visits (where they assist VEWs and AEOs in their field work and farm trials, respond to problems raised by farmers, and learn firsthand what is being done by farmers and extension in the field); and being trained themselves, mainly by research (particularly at a two-day monthly workshop with research staff and by visiting research facilities to maintain close contact with the latest research findings), and by conducting farm trials.

---

[1] Each country has a particular nomenclature for staff positions: local equivalents should be readily identified. In India, the Subdivisional Extension Officer is usually known as the Subdivisional Agricultural Officer, the District Extension Officer as the District Agricultural Officer, the Zonal Extension Officer as the Range or Division Joint Director of Agriculture, and the Director of Extension as the Additional Director of Agriculture (Extension).

This division of duties means that each fortnight a team of SMSs normally has four days available for fortnightly training sessions. Since the class size of the fortnightly sessions should not exceed about 30 persons, one team of SMSs can train a maximum of about 100 VEWs (since 100 VEWs normally require about 15 AEOs, who also are trained in the fortnightly training sessions). If there are more VEWs than this in a subdivision, additional teams of SMSs will be required. The Training Officer, in conjunction with the SDEO, is responsible for the logistics and administrative arrangements for each fortnightly training session and for all other in-service training activities in his subdivision. He is also responsible for providing training in extension methods to all staff, and for organizing and helping other SMSs to give short special training courses to extension staff of the subdivision.

District Level

The extension service in a district is under the control of the District Extension Officer (DEO). The DEO supervises the work of his SDEOs,

Skill practice in fortnightly training

Directorate of Agricultural Extension, Thailand

particularly by making field visits that coincide with their visits, in the course of which he meets VEWs, AEOs, and, most importantly, farmers. During these visits, he determines the impact and effectiveness of extension activities and of all his staff. The DEO should also regularly participate in training sessions of district staff. He must ensure that there is coordination between input and other agricultural support services and extension, and that close, effective links are established and maintained between extension and research.

The DEO is supported by a team of SMSs. District-level SMSs will backstop areas where specialized support is not yet required at the subdivisional level; these might include farm management, water management, farm women's activities, farm implements, animal husbandry, or crops of special local relevance. Among the district-level SMSs, there may also be more senior specialists in the major disciplines represented by the subdivisional SMSs. District SMSs have the same tripartite division of work responsibilities as their subdivisional colleagues: training others, making field visits, and being trained. They participate regularly in fortnightly training sessions, and provide technical support to subdivisional SMSs. It is also their responsibility to transfer know-how and feedback from one subdivision to another. District SMSs also have an important role in ensuring the active participation of subdivisional SMSs in monthly workshops and the overall quality and relevance of these workshops, in developing and overseeing the district farm trials program, and in conducting short in-service training courses for extension staff of the district.

Zone Level

Where there are too many districts (i.e., usually more than eight or so) for effective supervision by headquarters staff, an intermediate administrative level between headquarters and the district—the zone— is required. The Zonal Extension Officer (ZEO) is responsible for the smooth operation of all extension activities in the districts that comprise the zone. To do this, he undertakes field supervision work to see that the extension service is effectively reaching farmers with relevant technology and that farmers are provided with good advice and adopt extension's recommendations. The ZEO also regularly reviews the quality of training activities and the coordination of input supply and demand with extension activities and messages; he pays particular attention to the quality of monthly workshops as well as the relevance of preseasonal workshops where extension and research priorities and activities for the coming season are determined. The ZEO is assisted by a few administrative staff; in some circumstances, he may also have a small team of Subject Matter Specialists.

Headquarters Level

The organization of extension at headquarters differs according to the responsibilities and structure of the Department (or Ministry) of Agriculture. One basic requirement at the headquarters level in any organization is that responsibility for extension is separate from other tasks, especially agricultural supply and regulatory responsibilities and other departmental technical functions. Not only should the extension service operate as a separate entity within the department, but its staff, being devoted exclusively to extension work, should comprise a professional cadre that is recruited and promoted on the basis of objective, relevant criteria.

Rural extension center in Indonesia

The main responsibility of extension headquarters staff is to ensure the effective operation of the extension system throughout the state or country. Headquarters' attention should focus in particular on the quality of training and technical guidance received by all staff; coordination with and support from research and input and other agricultural support agencies; the promotion of broader agricultural and development activities; the quality and specialization of the service and its staff; exchanges of relevant experiences and ideas; monitoring and evaluation of extension activities; and the general administration of the extension service.

The Director of Extension, who reports to the Director of Agriculture, should be assisted by three deputies and a number of Subject Matter Specialists. One deputy should be in charge of administration, including personnel management and finance. A second should be responsible for all technical and professional aspects of the service and oversee the cell of senior Subject Matter Specialists. A third deputy should be responsible for the execution and implementation of the work at all levels, internal monitoring of the system, and advising the Director of Extension of steps required to increase the impact of the extension service on the farmers it serves. The SMSs at headquarters, who will include a Training Officer and represent the main disciplines of SMSs at lower levels as well as in other fields, are responsible for developing in-service training programs in their fields of specialization for extension personnel; maintaining close contact with research workers throughout the country and elsewhere; and providing technical support to other SMSs and extension staff and joint research/extension committees.

5. Other Operational Features

Agricultural extension and research are mutually dependent. Extension requires the findings of research to teach farmers, as well as the continuous support of research in solving farmers' problems. Without research's backing, extension may provide farmers with general support, timely reminders, and demonstrations of better practices, but it is unlikely that it can effectively transmit to farmers the significant improved practices (including new varieties, crops, and cropping patterns) that lead to the marked increases in productivity that are required for rapid, sustained agricultural development. Similarly, research requires extension's guidance on problems that farmers face and on which attention should be focused.

While strong links between extension and research are essential for either to operate effectively, they are often impaired by traditional institutional arrangements. Responsibility for agricultural research is often separated from the Agriculture Department. Extension and research need not be the responsibility of one organization, but there must be close contact and coordination between them. This is often difficult to attain. Not only are extension and research in many countries the responsibility of different organizations, but their headquarters may frequently be in different locations. Sometimes, extension and research even operate under different concepts of their purpose.

Links with Research

The links between extension and research should be strengthened by procedures that encourage systematic interaction between their staffs. These links must ensure that research is subject to continuous pressure from farmers and field staff, and that extension is up-to-date with significant research activities and findings. In training and visit extension, the most formal of these procedures are joint extension/research committees and workshops, where extension and research priorities are discussed, and production recommendations and applied and adaptive research programs are established and their results evaluated.

45

A most significant venue of research and extension linkages is the monthly workshop of extension Subject Matter Specialists (SMSs), District and Subdivisional Extension Officers, and research specialists (from regional research stations, or agricultural colleges, universities, or institutes). These workshops serve to train SMSs in the technology that farmers will need over the coming four to eight weeks. They also have the critical function of facilitating extension and research to meet frequently and regularly to undertake their joint responsibilities of developing production recommendations, reviewing farmers' production problems, planning farm trials, and analyzing and interpreting the results of such trials.

A second major, regular extension/research workshop is the zonal workshop held before each season with senior extension and research staff of a zone. (As with the monthly workshop and fortnightly training session, some farmers and representatives of input and marketing organizations also attend.) This workshop determines, for the season, the agricultural development objectives and strategy for the zone, general production recommendations to be promoted, applied and adaptive research activities, the farm trials program, and estimated input and market requirements. A similar meeting is also held before each season at the state or national level by senior extension and research staff.

Although formal extension/research committees and workshops provide one framework for extension/research interaction, effective linkages depend at least as much on a number of less formal measures. These include personal contacts and casual meetings between extension and research staff; exposure to one another during short training programs where extension staff are trained by researchers; occasional participation of research staff in fortnightly training sessions of extension staff; collaboration in farm trials; visits of extension staff to research stations; and joint visits of extension and research staff to farmers' fields.

Joint field visits are a particularly important means of extension/research contact. Research staff must make frequent field trips to review general field conditions and specific field problems raised by extension staff (usually, but not only, in monthly workshops), to inspect research plots and farm trials, and assess farmers' reactions to the recommendations they gave extension staff during monthly workshops. Extension staff should encourage such reviews by frequently inviting researchers to accompany them on their own visits, and by arranging for transport if this is a constraint to the mobility of research staff. Extension staff (especially SMSs) should often visit research stations to familiarize themselves with research activities and findings in their own particular areas of interest, and generally to update their technical knowledge.

46

Farm Trials

Trials carried out in farmers' fields—farm trials—are an important feature of extension operations. Such trials provide a final testing ground for research findings before they are recommended by extension on a large scale. They also provide a mechanism to facilitate close continuous working relationships between research and extension staff. The program of farm trials is developed jointly by researchers and extension personnel. Staff from local research stations and district level SMSs provide technical guidance to the subdivisional SMSs who, together with Agricultural Extension Officers (AEOs) and Village Extension Workers (VEWs), are responsible for carrying out the farm trials. Subject Matter Specialists and research workers together analyze the results of these trials.

Farm trials should be simple, normally involving no replications on the same farmer's field. The reliability of the results of these trials is ensured by carrying them out in relatively large numbers. Since some risk is involved, the plot size should not be too large. Compensation for land, inputs, labor, and so on is generally not provided to the farmer unless a completely new, unknown input or implement is required. The crop yield is usually not guaranteed. While the farmer faces some risk in a farm trial, the risk is small because, in order to be tested in such a trial, the recommendation has had to undergo considerable experimentation by research, during which its general feasibility has been established. Moreover, the trial takes place on only a small part of a farmer's field. The risk is further offset by the fact that the farmer will benefit from extra attention from extension personnel while the trial is being conducted, and, if successful, he will be in a better position than most farmers to adopt the practice more widely in following years.

Each VEW should be involved continuously in some farm trials under the guidance of his AEO and SMSs. Since all trials must be closely monitored, the VEW should undertake no more than two such trials in a season. If a trial is successful, the VEW should explain it to other farmers, using the field as a demonstration plot and the site of a field day.

Other Extension Methods

Extension methods in the training and visit extension system focus on the individual contact of the extension agent and the farmer during regular and fixed visits to farmers' fields. In addition to the individual communication and skill-teaching that the extension agent uses during these field visits, a number of other extension methods can increase his effectiveness in teaching recommended practices to farmers and in encouraging their use.

47

One important way is to use a farmer's own field, where he has implemented a particular recommended practice, as a demonstration field. It is more convincing if farmers see a practice successfully applied by another farmer on his own field and using his own resources, and have it explained to them by that farmer, than if they see a formal demonstration that has been prepared by government officers with departmental inputs. Such small-scale, individual demonstrations are built into the training and visit system, since the VEW normally recommends that a farmer adopt a newly recommended practice on only part of his field. This reduces the risk faced by a farmer, and enables him and others to see the difference between the recommended and usual practice on the same field.

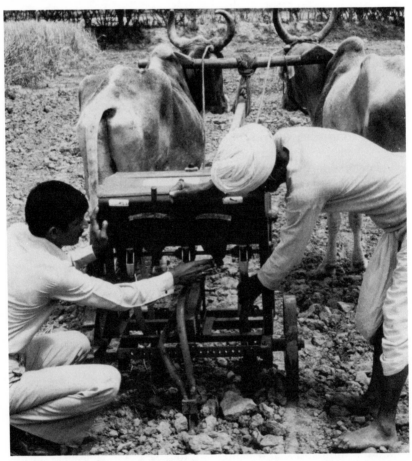

Learning how to calibrate seed-fertilizer drill

Directorate of Extension,
Government of India

From time to time, meetings with farmers are useful. For example, they may be held—on scheduled or extra visit days—before the start of a season to discuss the planned introduction of the reformed extension system, the extension program for the season, or to spread rapidly a recommendation during pest or disease emergencies. Meetings should, however, not be held at the cost of time during which the extension worker could be meeting farmers in their fields, and not as a matter of routine, particularly if attendance is slight or comprises chiefly those farmers who are also met in the fields. The purpose of field visits is to enable the extension worker to meet with, and teach recommendations to, contact and other farmers, and to learn of their production conditions and problems. Formal meetings should only be held so long as they clearly help in this. At times, it may be useful to show relevant charts, slides, and even short films at these meetings, and to distribute pamphlets on extension activities and recommended technology. Such support can be effective, but is not a substitute for field work.

Extension workers may organize field days for one or more farmers' groups to allow farmers to see plots where recommended practices have been adopted. AEOs, SMSs, Training Officers, and other extension staff, should participate actively in group activities organized by VEWs.

Mass media—newspapers, radio, and even television—can reinforce the work of extension by giving publicity about the system and by providing farmers with general technical advice. Relevant, attractively presented material (particularly broadcast over the radio), combined and coordinated with frequent field visits of extension agents, can ensure a quick, broad impact.

The extension agent should not work alone in deciding which extension methods to use. Guidance and support should come particularly from his Training Officer, but also from his AEO, Subdivisional Extension Officer, and SMSs. Most extension services have staff specifically responsible for information and publicity: These officers should work closely with field staff and SMSs to prepare a continuous stream of relevant material for the mass media, as well as visual aids for training and for field staff.

Other Training

Regular in-service training for staff is required in addition to the fortnightly training of VEWs and AEOs, monthly workshops for SMSs, Subdivisional and District Extension Officers, and research staff, and seasonal zonal meetings for senior extension and research staff. The training needs of staff vary greatly. The Director of Extension, assisted by a senior Training Officer, should annually determine the general training priorities. Taking account of the needs of individual staff, a

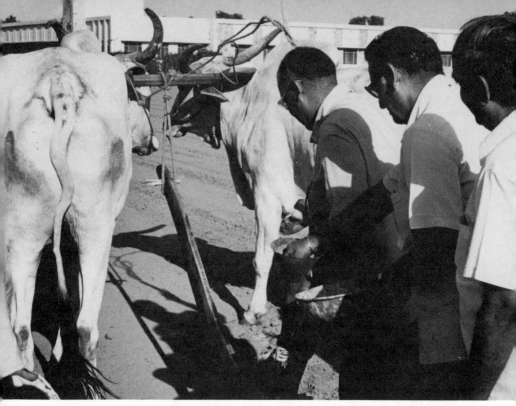

Monthly workshop participants practice skills in the field

B. Z. Mauthner

training program must be developed. As a rule of thumb, every member of the extension service should attend each year at least one special training course (i.e., training other than the regular sessions) and one short orientation or reorientation session. Many of these courses, especially those for VEWs and AEOs, may be arranged and given by local extension or research staff; others will be given by staff of agricultural training centers and universities.

A very important training requirement is thorough orientation and reorientation of all staff to the reformed professional extension system and, in particular, to the principles upon which it is based, the functions of different staff, working procedures, and the formulation of production recommendations and impact points the system is to convey. A main goal of such orientation training is to develop among staff of all levels a good understanding of the system and their role in it. Orientation training should be given continually, not only for staff new to the system, but also as refresher training for other staff.

All staff continue to require technical training to develop their understanding of key production recommendations for locally important crops, as well as to update their knowledge of general relevant agricultural technology. Such requirements of technical training for VEWs and

AEOs may be met through short courses (two days to two weeks in duration) given by SMSs or local research staff. Subject Matter Specialists and senior extension staff will require more intensive training, often at agricultural universities or central research institutes, and even overseas.

Since many extension staff are engaged in supervision and management, appropriate training should be given in these aspects. All staff, not only VEWs and AEOs, will require intensive training in extension methods and relevant communication techniques.

Training courses like those noted above are not meant to answer the academic needs of staff, but rather to upgrade the quality of staff at each level to allow them to contribute more effectively in an ever-changing agricultural and technological environment. For this reason, training needs must be identified locally and individually, and the courses must be practical. In addition to short or medium-term courses (which can be from two or three days to two or three months in duration), appropriate staff should also be enabled to attend courses of longer duration and even undergraduate or postgraduate studies.

Aside from formal training courses, informal means of training must not be ignored. Every contact between supervisors (including SMSs) and extension staff, particularly in the field, should be taken as an opportunity for guidance and strengthening. Also important are the opportunities given to extension staff, individually or in small groups, to review extension activities in other areas. If carefully planned and based on specific objectives, such visits (which may be overseas) quickly spread good extension methods.

Extension Coverage

An agricultural extension service organized on training and visit lines is likely to focus on field crops (and perhaps some tree crops) and, within these, on the few most important crops of farmers. Over time, however, the extension service should be expanded to cover most farm-based production activities for which farmers require technological guidance. Animal husbandry (particularly nutrition aspects), horticulture, and farm forestry are examples of areas usually not served at first by a professional extension service but which over time should be covered. To do this, the extension service will need to become increasingly specialized. Involvement of the extension service in ancillary farm-based production activities must be confined to technology transfer, the basic function of a professional extension service. The concerned technical department should continue to be responsible for policy decisions, input support, research guidance, general training, and so on.

51

The main changes required in the extension service to perform these additional tasks are providing for specialized SMSs and the reorganization of training activities (including participation by research) to cover the new areas, although the number of field staff may also need to be enhanced in local areas. Where SMSs who can help develop recommendations for, and give training in, the new areas of coverage are not available in the Department of Agriculture, the service of appropriate specialists from the concerned technical departments or elsewhere will have to be arranged. Participation of such complementary SMSs in seasonal and monthly workshops will sometimes be adequate, though it will often also be required in fortnightly training sessions, since it is there that extension field workers learn recommendations to teach farmers.

Where there is a marked seasonality of farm activities, the work programs of extension staff may be arranged so that they can devote time to these ancillary farm-based production activities in less agriculturally active periods. In addition to being useful for advising farmers

Extension workers can also deal with poultry

on these other farm-based activities, less busy periods for farmers have many other uses for extension staff. They can be used for field workers to receive additional training, or for them to give farmers more general training than is possible during the time available for visits in other seasons. They are also a good time for VEWs and AEOs to check on the diffusion of extension recommendations, to review with farmers their experiences of the previous season and their reaction to recommendations, and to assess priorities for the coming season. Used efficiently, such periods can give extension workers a good opportunity to strengthen their field activities.

Monitoring and Evaluation

Effective monitoring and evaluation contributes significantly to the quality of agricultural extension. It helps ensure that the extension service operates efficiently, enables management to take action to overcome shortcomings in extension operations, and provides policymakers with appropriate information on which to base decisions. Monitoring and evaluation is not a substitute for the direct field supervision that is an integral part of training and visit extension, but it is an independent and quantitative cross-check of the progress and results of field activities. It is not a faultfinding mechanism, but rather a supportive means to suggest problem areas that may not be readily apparent through the regular review of extension activities. Used positively and constructively, monitoring and evaluation can help create the atmosphere of trust, honesty, and self-criticism upon which good extension depends.

Monitoring keeps track of extension activities and of progress in the implementation of the extension system. Evaluation determines the impact of extension activities, particularly on the production and income of farmers. Monitoring and evaluation of training and visit extension involves mainly three activities, the first of which is being done by extension management and the other two by a discrete monitoring and evaluation unit. Extension management staff are responsible for the routine internal monitoring of extension activities (staff employed, training conducted, field visits made, and so on). Monitoring and evaluation unit staff conduct surveys of farmers covered by the extension service, focusing on visits made by extension staff, recommendations taught to farmers, and the impact on crop yields of these recommendations. The third monitoring and evaluation activity is the undertaking of special studies of particular aspects of the extension system by monitoring and evaluation unit staff or consultants.

Reporting of all monitoring and evaluation must be concise and quick if it is to be of use to extension management. To do their job effectively, staff of the monitoring and evaluation unit must be properly trained, be autonomous from the extension service, and have adequate mobility to conduct survey and other work as scheduled.

Incentives

Under the reorganized system, the entire extension staff and especially the VEWs will work much more intensively than before. Staff will be compensated for this in part by the increased job satisfaction—in the case of a VEW because he feels that he has been transformed into a respected member of the rural community. However, to attract the best possible candidates into extension work and to maintain the morale of the existing staff, salaries at all levels (and especially at the VEW level) need to be carefully reviewed. In most cases, it will be necessary that adjustments are made properly to reward work well done. Promotion policies should ensure that staff who perform well have a reasonable assurance of professional growth and financial gain. The establishment of a professional cadre of extension staff, and within this a cadre for SMSs and also for VEWs, will help regularize promotion criteria and prospects. In addition to adequate procedures for vertical promotion, salary scales of all staff should be reviewed and extended if necessary to provide adequate financial rewards within grades and to put extension personnel on a par with other similar staff. Adequate allowances must be established and maintained to enable and encourage extension staff to undertake the extensive, frequent travel that is required in their work. While periodic specialized training is a built-in part of the extension system, sometimes training—for example, university education and overseas courses and study tours—can also be used as a reward for able staff.

6. Personnel and Physical Requirements

The establishment of a professional agricultural extension service along the lines of the training and visit system will normally require additional personnel, training programs, vehicles, residential facilities, equipment, and financial resources. The system, however, readily incorporates available extension personnel and other resources, redeploying them in an effective manner. The costly duplication of special development schemes is eliminated. Staff in nonproductive jobs are reassigned. In this way, the incremental costs can usually be kept quite low, depending on the state of the existing service. It is necessary to review what is required, to evaluate the resources that are available at each level in the extension service, and to determine the difference.

Village Extension Worker—Circle Level

The key to determining the number of Village Extension Workers (VEWs) required—and hence of staff needed at all other levels—is the number of farm families the extension service is to serve. How many VEWs are required is determined by the total number of effective, operating farm families in the area to be covered, and the number of farm families one VEW can reasonably reach. Factors affecting the desirable ratio of VEWs to farm families are discussed in Chapter 4. Once the required number of VEWs is determined, the number of available VEWs is subtracted from the total required to obtain the number that needs to be recruited. If a fairly large number of additional VEWs is needed, and it will be some time before they can all be in position, then the area to be covered under the reformed extension system should be limited initially. The temptation to spread available staff thinly (by increasing the number of farm families served by a VEW) should be resisted, because no staff can work effectively nor make a significant impact if they are overburdened. Once an obvious impact is made in a limited area, there will be strong support for expanding the system as rapidly as possible.

In the long run, the aim of the extension service should be that all Village Extension Workers have a relevant university degree and a practical agricultural background. This is not easily achieved, however, particularly where there is considerable reassignment of staff to extension functions or where graduates are just not available. At the least, VEWs should be secondary-school graduates with practical agricultural backgrounds and some preservice training in agriculture. The system of regular fortnightly training sessions, backed by short special courses, as well as intensive guidance by Agricultural Extension Officers (AEOs) and others in the field, provides strong and continuous support for VEWs and can help overcome some initial shortfalls in staff qualifications.

Preservice training courses for VEWs—where the candidates are secondary-school graduates with agricultural backgrounds—should last about one year. This training is most effective when it is practically oriented and is interspersed with periods of regular work by the trainees as VEWs under the close guidance of experienced AEOs. Preservice training should focus on those aspects of agriculture that are relevant in the areas to which the VEWs will be assigned. It should have a considerable emphasis on extension methods and communication skills. A one-year preservice course might be split into six-month sections with a year or so of field assignment in-between. In this way, large numbers of VEWs can be put in the field quite rapidly and, at the same time, acquire a good mix of formal learning and practical experience.

In addition to preservice and routine fortnightly training, Village Extension Workers should occasionally receive training in specific relevant technical areas. Most of these special courses usually need not last longer than three days to a week or so. Each VEW should attend one special course (as well as a short orientation course) annually, for which appropriate organizational and financial arrangements will have to be made.

Each Village Extension Worker should have a bicycle or motorcycle to enable him to reach his farmers' groups easily. To ensure this, a government will normally need to provide loans on attractive terms. The bicycle or motorcycle will be owned by the VEW, and he will be responsible for its repair and maintenance. An adequate, fixed monthly travel allowance will be required to meet the cost of maintaining the bicycle or motorcycle and to cover the cost of meals taken in the field.

To ensure close contact with the farmers he serves, the VEW must live in his circle. Unless he already owns, or is provided with, quarters in his area of jurisdiction, he should rent quarters there. If housing is not provided, the VEW requires a housing allowance sufficient to cover full rental costs. Where houses are not available for rent, funds should be provided to construct a modest house compatible with both village standards and the position of the VEW.

In addition to the activities described above, funds should be budgeted for the VEW's extension operations, including funds for farm trials, field days, materials to make visual aids, the purchase of samples to show farmers, and so on.

Agricultural Extension Officer—Range Level

Since one Agricultural Extension Officer can effectively supervise only six to eight VEWs, the number of AEOs required is approximately one-eighth the number of VEWs. Staff at this level should be university graduates where possible, although practical field experience and organizational ability are as important in this position as academic qualifications. Agricultural Extension Officers should come from agricultural backgrounds and, preferably, have some experience with extension. VEWs with strong, technical field and supervisory skills, as well as appropriate educational qualifications, can make good AEOs, although not all good VEWs are suited to be AEOs given the significant differences in the responsibilities associated with the two positions.

Agricultural Extension Officers need to visit at least one or two VEWs in a day. Each AEO should, therefore, have access to a loan to purchase a motorcycle, which will belong to him and for the maintenance of which he will be responsible. To encourage and facilitate field travel, each AEO should receive a mileage allowance to cover costs of operating the motorcycle as well as a daily allowance to cover expenses in the field.

Senior staff regularly visit the field and meet with farmers—here, the Director of Agriculture, Kenya

Like VEWs, AEOs must live in their area of operation. If government quarters are not available in the range, AEOs should be provided with a housing allowance. Appropriate housing should be constructed as necessary and as funds allow. Funds will also be required for the in-service training of AEOs (they, too, should attend at least one special course annually) and for costs involved in their extension work (such as for training aids and field days).

Subdivision Level

At the subdivision level, one Subdivisional Extension Officer (SDEO) is required for about eight Agricultural Extension Officers. If there are more than eight AEOs in a subdivision, Assistant SDEOs should be appointed to maintain a ratio of approximately 1:8. The preservice qualifications of subdivisional officers are similar to those of AEOs. However, they require greater administrative ability and field experience since, in addition to supervising AEOs, the SDEOs are responsible for coordinating the work of subdivisional Subject Matter Specialists (SMSs) and for organizing and taking an active role in fortnightly training sessions and special training courses for AEOs and VEWs; they should also participate actively in monthly and seasonal zonal extension/research workshops.

Subject Matter Specialist (and Training Officer) positions are often difficult to fill, when an extension system is first organized on training and visit lines, because such a position rarely exists in the previous system. The basic qualification of an SMS is that he should know his subject well and be reasonably good at training. It may be best at first not to impose fixed rules for qualifications. Although, in unusual cases, a high-school graduate with substantial experience may make a good Subject Matter Specialist, normally the SMS should be a college graduate (a relevant M.Sc. is usually best) with extensive practical experience in agriculture and specialization in a chosen field. Such people are usually in short supply, and so a start should be made with existing staff. With special intensive courses, regular monthly and seasonal workshops, a demanding training load in fortnightly training sessions, and heavy exposure to the field and to research, staff without all the required formal qualifications often become very effective Subject Matter Specialists.

The term "Subject Matter Specialist" is relative; in the initial stages, the areas of specialization can be fairly broad. A careful search of the existing extension service and of special agricultural schemes usually reveals a reasonable number of staff of sufficient specialization to qualify as SMSs and Training Officers. By using such personnel—bolstered by short, intensive courses—it is possible to make a start rather than waiting indefinitely for more suitably qualified candidates. As the standards of farming improve and agriculture becomes more sophisticated, SMSs will need to become more narrowly specialized.

Usually, too few good SMSs are available to start with, so they must be used economically. At the subdivision level, a ratio of one SMS team (consisting usually of at least an agronomist, a plant protection specialist, and a Training Officer) to about one hundred VEWs is recommended. Since the number of SMSs required is relatively small, it is usually fairly simple to arrange in-service training programs for them. Funds should be available for the training of SMSs and Training Officers within the country and, if appropriate, elsewhere.

Subdivision staff need to be highly mobile. Subdivisional Extension Officers, Assistant SDEOs, Subject Matter Specialists, and Training Officers have to cover a large area regularly. They travel to the VEW training centers, farmers' fields, and research stations. For this reason, there should normally be a vehicle for each SDEO and Assistant SDEO, and for each SMS team. These vehicles should be government owned and maintained. Subdivision staff should also be eligible for loans to purchase motorcycles, if they will use them for their work. Travel allowances that cover meals and other incidental field costs are required. Arrangements for housing and office space should be made; additional houses may need to be constructed, since subdivisional staff, like all extension staff, must reside in their area of jurisdiction. Subdivisional staff, again like all others, should have an adequate budget for farm trials, training aids, field days, and all other extension operations.

District Level

Staff at the district level both supervise extension operations at lower levels and provide technical and administrative support for the subdivision offices (usually three to five) of the district. In addition to the District Extension Officer, each district should have a team of Subject Matter Specialists. These SMSs should be more experienced and skilled than subdivisional SMSs, although their selection initially is likely to be compromised in a way similar to that at the subdivisional level. The district SMS team, which would normally comprise a few SMSs and a Training Officer, should back up the main specializations handled by subdivisional SMSs, as well as represent locally important subjects that are not covered at the subdivision level. Examples might include SMSs for crops of special importance to the area, farm implements, farm women's activities, or water management.

At the district (or at least zone) level, as well as at extension headquarters, there should be a well-equipped Information Unit that collects, prepares, and circulates—both within and outside the exten-

Farmer is being shown how to top-dress correctly

**Directorate of Extension,
Government of India**

sion service—relevant information, publications, and training aids. Staff of the Information Unit should prepare leaflets, charts, and other training aids and circulate information on the extension service's activities and recommendations. Particularly at headquarters, the Information Unit should publish a small periodical highlighting technical discoveries and the results of particularly successful extension work, as well as generally serve as a forum for exchanging ideas within the extension service.

Zone Level

Zone-level staff—which are needed when the number of districts is too large to be supervised directly from headquarters—comprise the Zonal Extension Officer and possibly a team of Subject Matter Specialists. As at the district and headquarters levels, appropriate support for office accommodation, transport, training, field activities, operating costs, and housing, must be provided to the zonal office.

Headquarters Level

Staff requirements at headquarters are minimal. Here, as throughout the extension service, priority is given to field-based, mobile staff over office-oriented staff. The fact that the reformed extension system needs few written reports minimizes office-staff requirements. In addition to the Director of Extension, no more than three additional senior staff members and a small team of Subject Matter Specialists are normally needed. The functions of these officers are noted in Chapter 4.

Provision should be made at headquarters for sufficient vehicles and travel allowances to enable headquarters staff to visit the field as frequently as their jobs entail. Staff and logistical support may also be required for carrying out and analyzing crop-cutting experiments to evaluate the impact of extension.

Training programs are needed at headquarters level both to keep the technical specialists up-to-date in their own fields and in extension methods, as well as to improve the administrative and management capability of the Department of Agriculture.

Estimating the Requirements

The various requirements of personnel, training programs, housing, vehicles, equipment, and extension operations can be readily estimated using the norms described above. The cost of organizing the extension service can (and should) be kept low by using existing extension staff   61 and all available other resources to the fullest extent possible. Utmost care should be taken that all staff and other resources, including those involved in special schemes related to extension, are appropriately blended into the reorganized unified extension service.

Once the staff and other resources available for assignment to the reformed extension system are known (and consequently also the incremental requirements), a phased schedule for the introduction of professional extension should be worked out. This schedule should take into account the rate at which new staff can be recruited, oriented, and properly trained. Unit costs of the elements required at each level of the new extension service should be estimated. By applying these to the estimate of phased requirements, a yearly budget for extension reform can easily be calculated.

7. Impact of Effective Extension

The extension system described in the preceding chapters clearly has the potential of becoming a powerful communication tool. It enables vast numbers of farmers to be reached very quickly. All contact farmers and a significant number of other farmers are in direct contact with trained, competent extension agents once every two weeks. Many more farmers are affected indirectly. The latest techniques can be taught to Village Extension Workers (VEWs) and Agricultural Extension Officers (AEOs) in a fortnightly training session, and, within two weeks, a substantial proportion of all farmers in the area served by the extension service will also have received the new information.

The potential of the system is clear. But what has been its actual impact? As the objective of agricultural extension is to increase farmers' production and incomes, its impact must be measured in the field. Impact may be evaluated from the quantitative effect of such indicators as yield and areas planted in accordance with recommended practices; the impressions of visitors to the fields where the system is operating; farmers' reactions to the new system; and the reactions of extension staff themselves to their new mode of work. The following paragraphs summarize the results of each of these ways of assessing the system, particularly with reference to India where greatest experience with the system is available. There is, however, no substitute for actually visiting an area where the system is operating.

Quantitative Impact

Assessing the quantitative impact of extension is difficult. Agriculture is a very complex activity with many interacting factors making it virtually impossible to determine with precision what part of any increase in production is due to which variable. It is particularly difficult to adjust for variations in weather, and this complicates comparisons between years. Differences between farmers (those covered by extension versus those not) during the same year can be questioned on the grounds that perhaps only the better farmers (or lands) were covered by the extension service and that these better farmers (or lands) would have had higher yields anyway. If yield increases are the result of a combination of extension and other things, what is extension's share? Attributing a particular share of production increases to extension is difficult. It is not so difficult, however, to identify particular practices that farmers are pursuing after having learned them from extension— or from other farmers who did so. The point is that extension plays a major role in farmers' decisions on whether and how to use agricultural inputs: increased production as a result, for example, of better fertilizer use by farmers on the advice of extension is clearly due both to the input—fertilizer—and to extension.

Even the most carefully designed and executed evaluation can only minimize such problems; it cannot avoid them. Studies of this kind are very useful and have been initiated in several countries by agencies separate from the extension service. A detailed, rigorous evaluation of the system is beyond the scope of this booklet. However, preliminary estimates of the impact of extension have been made in some areas. Most of these studies have been carried out by the statistical branch of the Department of Agriculture or the monitoring and evaluation cell set up under the reformed extension system; some were done by separate agencies. The studies are usually based on crop-cutting experiments of samples of farmers, according to standard statistical procedures. Although the interpretation of these results is subject to the problems outlined above, steps have been taken to reduce such interference. As an indication of the impact of the extension system, examples follow from a number of states in India.

The agricultural extension system of Gujarat was reformed along training and visit lines with effect from 1979–80. A number of significant changes in agriculture have occurred in Gujarat since that time. Perhaps most striking has been the breakthrough in groundnut production. The extension service actively advocated the sowing of groundnut in the premonsoon and summer seasons by making better use of irrigation resources, improved seed varieties, and appropriate plant protection measures. (These extension activities complemented a number of financial incentives and corresponded with favorable market conditions.) From 25,500 hectares in 1978–79, the area under summer groundnut increased to 174,000 hectares in 1981–82, and was projected to be 225,000 hectares in 1982–83. In contrast to the expansion of the groundnut area, the area under cotton decreased from an average of 1.8 million hectares during the four years prior to 1979–80 to an average of 1.6 million hectares over the period since 1979–80, but production actually improved slightly with average yields per hectare increasing from 181 kilograms to 198 kilograms. Cotton growers also benefited from education in integrated pest management; many farmers halved the number of sprays given to cotton.

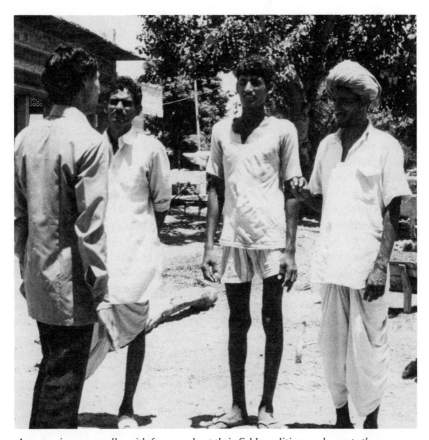

An extension agent talks with farmers about their field conditions as he meets them

B. Z. Mauthner

Yields of other major crops have increased, many significantly, over the period since the introduction of professional extension in Gujarat. One innovative measure consisted of promoting the growing of sorghum during the winter season (when its yields were almost double those obtained during the traditional monsoon season). Improvements occurring parallel with this increased production have been that the state agroindustry corporation has further developed its distribution of fertilizers and implements, the seed corporation has improved the production and distribution of better seed, and the marketing cooperative has developed a marketing system for groundnut. This is not to say that these changes came about exclusively because of extension. However, it is unlikely they would have occurred without the dynamism agriculture has displayed since the introduction of the extension system and without the demand generated by it at the farm level.   65

In Karnataka, the extension service was reorganized at the same time as in Gujarat, though its introduction over the state was phased over a three-year period. Attention, among other things, has focused on the introduction of new crops and improved varieties, as well as on basic fertilizer applications. Red gram was introduced in the southern areas of the state; within three years, 28,000 hectares were being planted annually. Green gram grown as a catch crop is another extension-promoted innovation. And improved varieties of rice and millet and a short-duration cowpea have been introduced over large areas. The extension service effectively promoted the use of gypsum on groundnut, rhizobium culture on pulses, and di-ammonium phosphate (DAP) for pulses and oilseeds. In 1980–81, which turned out to be a severe drought year, DAP was used on some 160,000 hectares.

The extension system was introduced in Maharashtra state only in 1981–82, to be implemented over a three-year period. The system quickly made an impact in the districts where it had been first started. In fact, the significant impact of the new system resulted in farmers and interested persons in other districts to press for its early introduction; the proposed three-year phasing was reduced to two years. The reformed extension service has put initial emphasis on suitable improved varieties, adjustments in cropping patterns, and appropriate pest control measures. In a number of districts, the area under hybrid sorghum has increased three times to four times (e.g., to 23,000 hectares in Solapur district), with yields approximately double those of local varieties. Intercropping of pulses with cotton and sorghum, and of soybean with cotton, has been successfully promoted. Elsewhere, under extension's guidance, farmers are economically growing safflower as a single crop where previously it was sparsely grown as an intercrop in sorghum. The area under summer groundnut has increased significantly, and the number of sprays given cotton has been reduced in line with recommendations on integrated pest control taught by extension workers.

While the impact of the reformed extension system is by no means even, similar examples may be cited from each area and country where it has been adopted. Statistical analysis does not permit a judgment on the extent to which increases in production and yield and technological improvements are due to extension alone and how much to other inputs or investments. In each of these cases, however, inputs were generally available before the introduction of the reformed extension service but were not widely used. Nor did the availability of irrigation increase substantially. In every case, the major change was in the quality and professionalism of the extension service itself, strongly suggesting that extension had a major catalytic effect on boosting production and yields.

Visual Impact

A visit to an area where professional extension is working provides clear visible evidence of the impact of extension. This is the same kind of impact that farmers experience and that convinces them to follow extension's recommendations. It is why extension workers encourage farmers to adopt recommended practices on only part of their fields: if the practice is worthwhile, the results speak for themselves, and the farmers who tried the practice will continue it in following seasons, while others will also adopt it. The visibility of the effects of extension and the speed at which they occur amaze most visitors. The conditions of the fields in many areas provide evidence of sound agricultural practices. Weeds are few, plant populations are at, or approach, optimal, sowing is in lines, and pests are well controlled—all clearly showing a basic transformation in agricultural practices and, hence, in yields and incomes.

There is often also a strong suggestion that the increased incomes are fostering development of an area in a broader sense. Better food is consumed, more children are sent to school, and more and better houses are being constructed, giving additional employment opportunities. As a result, labor wages in many of these areas have risen. Such signs of progress are a reliable indication of a major improvement in living standards.

Line transplanting in Thailand

Farmers' Reactions

Farmers' reactions to this professional extension approach, where it has been implemented well, are generally enthusiastic. The frequency and reliability of the visits of the Village Extension Worker at known intervals, combined with the soundness of his recommendations and advice, induces an almost immediate positive response. The VEW in many areas is virtually the only government officer who regularly visits, and spends time with, farmers, teaching them technology that can quickly increase their income. After only a few rounds of visits, when the function and utility of the VEW has become apparent, he is often approached for extra advice between visits; his fixed visit day becomes an important element in farmers' lives.

It is not unusual for farmers to seek a reversal of transfer orders for their VEW, and to arrange accommodation and other assistance to ensure he keeps coming to their village. Once farmers have become used to good service from a VEW, they are keen for such assistance to continue. With their knowledge of the fixed visit day, when the VEW has no other responsibility but to visit them, farmers become agitated if he does not visit as expected, and often report the matter to the VEW's supervisor, the Agricultural Extension Officer.

Reaction of Extension Personnel

Perhaps the most convincing testimony to the effectiveness of the professional extension approach is the reaction of the field-level extension staff. These people are a dispirited group in many countries. They have an enormous task but are rarely provided with even the minimum required administrative and technical support. Over the years, many have become quite cynical; they have learned by experience that it does not pay to do a good job, and that their knowledge is inadequate to help farmers effectively. Many extension services and even entire departments of agriculture are a catch-all of numerous and dubious rural development programs. Agriculture is not regarded as a specialized field or as a profession; without this recognition, it is usually left without an effective organization, and so cannot operate like a profession: it is a vicious circle.

Where extension has been reorganized on the principles described here, the situation has changed. Given an achievable task and the means, time, and training to do it, most VEWs, AEOs, and others have responded sincerely and eagerly. Suddenly, they can see clear results from their work after years of achieving little. They begin to have pride in their work and role as professional extension workers, to gain the confidence of farmers, and to become respected figures in the local community. Other staff in the extension service have shown a similar positive reaction to the system. Regular training, definite objectives, and fixed and realistic work programs, close links with research and with farmers, and the abandonment of vague, multifarious nonextension responsibilities all combine to give staff of the extension service—and, ultimately, of the Department of Agriculture—a professional, purposeful image. Not only does extension reform bring about a change in how the service sees itself, but its apparent sense of purpose, systematic links with farmers and research, and efficient organization enable others also to see it as a professional and competent force. This, in turn, often changes people's perceptions of the Department of Agriculture, which becomes to be seen as a strong and effective force to facilitate rapid development.

Another significant impact of the system is its contribution to the generation of practical agricultural research and establishing a means of systematic and direct communication between farmers, government (in the guise of the Department of Agriculture), and research. Although not all agricultural research can or should have immediate benefits for farmers, the links between extension and research staff, particularly through monthly and seasonal workshops and joint field visits, ensure

Farmer shows off a good cotton field in India

that research remains continually aware of key constraints facing farmers, and that relevant research findings are quickly passed on to farmers. The fixed visit schedules of extension staff to farmers' fields, and their carefully delineated areas of operation, mean that virtually all farmers can be made aware of new technical advice very quickly. Similarly, production, marketing, or supply problems faced by farmers, can be brought to the attention of sufficiently senior government officers by VEWs during their fortnightly training sessions or even earlier. Effective and regular feedback from the field ensures that farmers' priorities and problems are considered in determining the technical advice they receive, as well as in motivating appropriate research and other agricultural support activities.

Priorities

The data and descriptions presented above all suggest that a properly organized extension service can have a major impact in a very short time, although it takes not less than ten to fifteen years of continuous efforts to have a well-organized and smoothly functioning professional extension service in place. In countries where the standard of cultivation and productivity is quite low, it is difficult to imagine a more productive investment. The incremental costs are minor both in relative and absolute terms, generally about $1 per hectare a year. The benefits are disproportionately large: additional yields of 0.5 ton to 1 ton of rice, worth $60 to $120 per hectare, is approximately what has been achieved in several areas where professional extension is operating. Where more than one crop is grown in a year, the impact is still greater.

It takes no sophisticated analysis to see that professional extension is a very good buy indeed. This does not imply that one should invest in extension only, instead of projects to develop, for example, input supply, agricultural credit, and water resources. But as a matter of priority, it seems logical initially to emphasize extension, because it is a prerequisite to reap fully the benefits of these other investments.

Professional extension costs little, achieves much, gives farmers and extension personnel alike self-confidence and pride in their work, rapidly generates increased production, and creates (and transmits) demands for other inputs and services at the farm level. Once this process is under way, it can provide the moving force that continuously facilitates further generation of new agricultural technology and development. Professional extension, tuned to farmers' needs and a country's capacities, is a most powerful tool to attain an early impact on productivity and farmer's incomes and, therefore, to improve the quality of life of millions of people on the land.

70

ANNEX

Main Features of the Training and Visit System of
Agricultural Extension

Annex

Main Features of the Training and Visit System of Agricultural Extension

Introductory Summaries of the Chapters of
Training and Visit Extension
by Daniel Benor and Michael Baxter (The World Bank, 1984)

1. *Introduction*

2. *Some Key Features of the Training and Visit (T&V) System of Agricultural Extension*

The training and visit (T&V) system of agricultural extension aims at building a professional extension service that is capable of assisting farmers in raising production and increasing incomes and of providing appropriate support for agricultural development. The system has been widely adopted in many countries. Considerable variation in the system exists within and between different countries, reflecting particular agroecological conditions, socioeconomic environments, and administrative structures. To be successful, the training and visit system must be adapted to fit local conditions. Certain features of the system, however, cannot be changed significantly without adversely affecting its operation. These features include professionalism, a single line of command, concentration of effort, time-bound work, field and farmer orientation, regular and continuous training, and close linkages with research.

3. *Role of the Village Extension Worker*

The Village Extension Worker (VEW) is the only extension worker who teaches production recommendations to farmers. He is just as specialized and professional as other extension workers. The responsi- 73 bility of all other extension staff is ultimately to make the VEW more effective in his work. The task of teaching farmers suitable technical practices and convincing farmers to try them is not easy. Hence, the VEW must receive intense support and guidance, and must not be burdened with nonextension functions. Moreover, the nature of his work and his achievements must be recognized personally and in terms of opportunities for professional growth and technical upgrading. The main responsibility of the VEW is to visit regularly each of the eight farmers' groups of his area of jurisdiction (the "circle"), and to teach and try to convince farmers to adopt recommended production practices. He must also advise farmers on the price and availability of necessary inputs and market conditions. He should report farmer response to recommendations, production problems, input demand and availability, and market conditions to his supervisor (the Agricultural Extension Officer) and in training. Days without a regularly scheduled visit or training are used for makeup visits, farm trials, and field days. In addition to making field visits for at least eight days, each fortnight the VEW must attend a fortnightly training session given by Subject Matter Specialists (SMSs) and a review meeting with his Agricultural Extension Officer (AEO).

4. Role of the Agricultural Extension Officer

The importance of the Agricultural Extension Officer (AEO) in good agricultural extension is frequently underestimated. The AEO has two basic functions. The first is to review and assist in the organizational aspects of the job of the Village Extension Worker (VEW); the second, to provide technical support to the VEW, in particular to see that production recommendations are effectively taught to farmers and that field problems encountered by a VEW, and which he himself cannot resolve, are passed on immediately to appropriate authorities. Like the VEW, the AEO is primarily a field worker. He spends at least eight days each fortnight in the field visiting each of the eight or so VEWs of his area of jurisdiction (the "range"), in particular to make sure that farmers are being visited regularly by the VEW, and that the recommendations they receive are appropriate and are adopted. He reviews whether contact farmers have been correctly selected, farmers' groups are properly delineated, and all farmers are aware of the VEW's visit schedule and activities. The AEO should also conduct some farm trials in farmers' fields, participate in fortnightly training sessions, and hold a fortnightly review meeting with his VEWs. In addition to these specific tasks, the AEO should take any steps that may be necessary to fulfill his main responsibility of helping the VEW increase his effectiveness as an extension worker.

5. *Role of the Subdivisional Extension Officer*

The Subdivisional Extension Officer (SDEO) has overall responsibility for effective agricultural extension in his subdivision. Through leadership, planning, and supervision, he must ensure that extension has a significant impact on agricultural production and farmers' incomes. To do this, the SDEO must use his initiative to take any action required to increase the effectiveness of the extension service. The SDEO is active in two main areas—field visits and training—in addition to coordinating information on the actual and likely supply and demand of agricultural inputs and on market conditions in his subdivision. He makes field visits on at least three days each week to review both technical and organizational aspects of the work of extension staff in his subdivision. He is the organizer, convenor, and leader of fortnightly training sessions. Monthly and zonal workshops and other extension/research meetings are attended by the SDEO mainly to ensure that significant relevant local conditions are taken into account in the formulation of recommendations and research activities. The SDEO should also ensure that extension staff of his subdivision receive adequate and appropriate special training.

6. *Role of the Subject Matter Specialist*

The Subject Matter Specialist (SMS) provides technical training and guidance to extension workers, has an important role in the formulation of production recommendations, and is a focus of links between extension and research. An extension service's ever-increasing requirement of technical specialization is met primarily through increasing the number of SMSs and improving their degree of specialization. SMSs are usually present at three levels (subdivision, district, and headquarters), at each of which they have three common functions: to make field visits, to train extension staff, and to be trained by and exposed to research. Approximately equal time is devoted to each function. In field visits, their main concerns are the correctness of recommendations taught farmers by Village Extension Workers (VEWs), farmers' reactions to production recommendations, and aspects of agriculture that require additional recommendations or referral to research. SMSs are the trainers at fortnightly training sessions as well as for some specialized short courses. SMSs are trained by research staff at monthly workshops and elsewhere. They also visit research stations, attend specialized training courses given by research, and conduct farm trials. This regular contact with research helps ensure that proposed recommendations take account of local production conditions. To perform their vital function effectively, SMSs should be hired, trained, and promoted within a specialized staff cadre.

7. Village Extension Worker Circles, Farmers' Groups, and Agricultural Extension Officer Ranges

Effective agricultural extension depends on extension messages reaching many farmers, and farmers' problems reaching extension staff, quickly and regularly. A key means toward this end are regular, fixed visits made by extension workers to specific groups of farmers within a precisely defined area. The groups of farmers for which the base-level extension worker—Village Extension Worker (VEW)—is responsible comprise the VEW "circle." The size of the circle is derived from a broad ratio of effective operating farm families to a VEW, which is based mainly on the compactness of settlement, the ease of communications, and the intensity of agriculture. There is no standard rule for determining the number of operating farm families to be served by one VEW. The number should not be so large that the VEW's messages are unable to reach quickly most farmers. A common ratio is one VEW to about 800 operating farm families. The farmers of a circle are divided into eight "farmers' groups," each of which will be visited on a fixed day by the VEW. VEW circles that are the responsibility of one Agricultural Extension Officer (AEO), the immediate supervisor of the VEW, comprise the AEO "range." A range should be sufficiently small (usually comprising not more than eight VEW circles) so that each VEW in it can closely and effectively be guided by the AEO. The areas of circles, farmers' groups, and ranges must be compact and contiguous.

8. Contact Farmers

Frequent contact between a Village Extension Worker (VEW) and all farmers in his circle is not possible. Instead, while being responsible to all farmers, on each fortnightly visit the VEW focuses on a small, selected number of farmers—"contact farmers"—in each farmers' group, and meets with any other farmers who are willing and interested to attend his visits and seek his advice. Contact farmers are identified by the VEW and the Agricultural Extension Officer (AEO) with assistance of the local villagers, especially village elders. Contact farmers are selected according to the following characteristics: (1) they should represent proportionately the main socioeconomic and farming conditions of their group and be regarded by other farmers as able and worthy of imitation: (2) they should be practicing farmers; (3) they should be willing to adopt relevant recommendations on at least a part of their land, allow other farmers to observe the practices, and explain the practices to them; (4) as far as size and composition of farmers' groups permit, they should come from different families; and (5) their farms should be dispersed throughout the group area. Tenants, sharecroppers, young farmers, and women farmers may be contact farmers if they possess these characteristics. No major type of farmer should be over- or underrepresented among the contact farmers of a group. Once a contact farmer becomes disinterested in the work of the VEW or becomes in other ways ineffective, he should be replaced.

9. *Visits*

A key feature of the training and visit system of agricultural extension are the regularly scheduled visits to farmers' fields by extension staff. Visits are made, on the one hand, to advise and teach farmers recommendations on relevant agricultural technology and to encourage them to adopt these and, on the other, to establish in extension and research an awareness of actual farmer conditions and needs. All extension staff on field visits should listen as much as they talk. The basic extension worker is the Village Extension Worker (VEW), who visits each of his eight groups of farmers on a fixed day once every fortnight. His visits must be regular, specific, and purposeful. On a visit, a VEW should teach production recommendations to as many farmers as possible, and certainly to all contact farmers, and attempt to convince them to adopt the recommendations on at least a small part of their land. The Agricultural Extension Officer (AEO) visits his eight or so VEWs regularly in the field—not less than once each fortnight—guiding, supervising, and giving technical support. The Subdivisional Extension Officer (SDEO) makes field visits for at least three days a week, providing support for VEWs and AEOs in technical and organizational aspects of extension. The SDEO is responsible for the effectiveness of extension in his areas, seeing to it that VEWs and AEOs work as required. Subject Matter Specialists (SMSs) spend one-third of their time in the field providing technical support to VEWs and AEOs. Field visits by AEOs, SDEOs, and SMSs are an important means of support for VEWs.

10. *Monthly Workshops*

The monthly workshop is the main venue of in-service training for Subject Matter Specialists (SMSs) and of regular contact between extension and research workers. A main purpose of the two-day workshop is to build up the technical skills of SMSs regularly in the field of their specialization, so they can meet effectively the actual technological needs of farmers. Another purpose is for researchers and SMSs to discuss and formulate relevant production recommendations for subsequent transferral to Village Extension Workers (VEWs) and Agricultural Extension Officers (AEOs) by SMSs at the next two fortnightly

77

training sessions. To be effective, monthly workshops must have a strong practical orientation and encourage discussion among participants. The learning that takes place through extension and research workers' discussing each other's experiences is as important as the formal learning of recommendations and solutions to farmers' problems. Monthly workshops should be held at a district level, if possible. They require considerable advance planning. Monthly workshops may be organized in a variety of ways, although there are basic activities that must be covered in each.

11. *Fortnightly Training*

The chief means of continuously upgrading and updating the professional skills of Village Extension Workers (VEWs) and Agricultural Extension Officers (AEOs) is the fortnightly training session, which is held for one full day each fortnight. At fortnightly training, VEWs and AEOs review farmers' reactions to previous recommendations, are taught specific recommended practices that will be taught to farmers during the coming two weeks, report field problems or conditions that need to be taken into account in these recommendations or which are to be passed on to research for investigation, and discuss, and learn from, each other's experience. The organizer of the training session is the Subdivisional Extension Officer (SDEO), helped by the Training Officer; the trainers are primarily subdivisional Subject Matter Specialists (SMSs). No more than about thirty VEWs and AEOs should attend a fortnightly training session. Representatives of local input and marketing organizations should also attend, and some local farmers may be invited. Training should emphasize a small number of production recommendations and impact points and encourage practical work by VEWs and AEOs. Fortnightly training sessions can be organized in a number of ways, although each should include some common activities and involve approximately equal time in teaching and in practical work.

12. *Production Recommendations*

Production recommendations are the specific agricultural practices that extension teaches farmers. They represent the most suitable and economically viable production technology for a crop under a farmer's production conditions. Without production recommendations, it is impossible to plan, implement, monitor, or evaluate extension work, and extension staff are unlikely to be able to assume their desired active, diagnostic role. All major crops and practices should be covered by production recommendations, although at any given time extension will emphasize selected key points from the recommendations. Recommendations should be so designed that farmers would be willing and able to follow them. Therefore, recommendations that represent a new practice for a farmer must be financially feasible, result in increased production and income, and entail minimum risk. Recommendations are provisionally developed at seasonal zonal workshops of extension and research staff; they are refined and modified at monthly workshops and fortnightly training sessions to take account of local field and production conditions. To be useful to farmers, recommendations must be continually reviewed and adjusted in light of changing production conditions.

13. *Linkages between Extension and Research*

Extension and research are dependent on one another for their successful operation. Extension needs research's findings and its solutions to technical problems to teach to farmers as production recommendations. Extension should serve as a main source for research to develop an orientation to, and maintain an awareness of, actual farm problems. While close linkages between extension and research are a necessity, they are not easy to achieve. Under the training and visit system of extension, systematic procedures have been established to promote and strengthen the necessary linkages by means of periodic meetings of research and extension staff in the monthly workshop, as well as through seasonal zonal workshops and the State Technical Committee. Through its frequency, activities, and composition, the monthly workshop is the most important of these meetings. Research/ extension linkages are also promoted through the training of extension staff by research staff, by collaboration in farm trials, and through visits of research staff to farmers' fields and of extension staff to research facilities.

14. *Applied and Adaptive Research*

Applied research is the development of new technology and its verification under different agroecological conditions. In many areas, applied research is the responsibility of agricultural universities or other research organizations rather than the Department of Agriculture. Adaptive research—which is referred to as "farm trials" in order better to represent its nature—is the adaptation of general recommendations to specific farming situations, in particular to farmers' resources and abilities, cropping patterns, and actual farm conditions. Adaptive research is usually the responsibility of the Department of Agriculture, 79 although farm trials are planned by extension and research together, executed by extension, and analyzed by extension and research. Subject Matter Specialists (SMSs), Agricultural Extension Officers (AEOs), and Village Extension Workers (VEWs) are all closely involved in farm trials, which are also reviewed in seasonal and monthly extension/research workshops.

15. *Supervision*

Supervision alone cannot produce good agricultural extension but good extension is rarely possible without effective supervision. Supervision should be tailored to fit the training and visit system of agricultural extension. It must not be based on paperwork or report writing. As supervision takes place at the location of the activity to be supervised, extension supervision is mostly conducted in the field, the only exception being supervision of training. Supervision must be thoroughly planned and thoughtfully implemented. The objective of supervision is to guide staff and help them become more effective, not merely to check whether they are doing their assigned tasks as required. Aside from Village Extension Workers (VEWs), all extension staff have some supervisory functions, the nature and intensity of which vary according to their level of responsibility. Extension supervisors, at all levels, should check (among other things) the end result of extension's work: whether farmers benefit economically and otherwise from extension.

16. *Diaries of Village Extension Workers and Agricultural Extension Officers*

The only written report required of Village Extension Workers (VEWs) and Agricultural Extension Officers (AEOs) is the daily diary. Properly used, a well-designed diary can greatly enhance the effectiveness of extension work. Diaries are used to record three main items: (1) basic information about the VEW's circle or AEO's range; (2) the extension worker's daily activities and problems encountered in the field; and (3) the main points discussed in each fortnightly training session and AEO/VEW fortnightly meeting. With such information, the diary serves as a guide to VEWs and AEOs in their field work and training, and to supervisory staff in their guidance to extension workers and to the problems they encounter in the field. It should not, however, be used to monitor or evaluate an officer's work. A diary should be relatively small and sturdy, so that it can withstand constant use in the field (where problems and observations are recorded as they arise). It should be available for perusal by officers, who should write substantive comments in it whenever they visit staff in the field. No copies should be made of the diaries and, of course, no copies should be sent to any officer.

17. *Monitoring and Evaluation*

Monitoring and evaluation is a management tool that can contribute significantly to effective extension. Monitoring keeps track of extension activities and progress in the implementation of the extension system. Evaluation determines the impact of extension activities, particularly on the production and income of farmers. Monitoring and evaluation of training and visit extension entails three functions, the first being done by extension management and the other two by a monitoring and evaluation unit: (1) routine monitoring of extension activities and their impact (staff employed, training conducted, field visits made, and so on); (2) monitoring and evaluation surveys of farmers covered by the extension service, focusing on visits made by extension staff, recommendations taught to farmers, and crop yields; and (3) special studies of particular aspects of the extension system. Reporting of all monitoring and evaluation must be done concisely and quickly if it is to be of use to management. To do their job effectively, monitoring and evaluation staff must be properly trained, be autonomous from the extension service, and have adequate mobility to conduct survey and other work as scheduled.

18. *Planning Extension Activities*

The organizational structure and fixed schedule of activities of the training and visit system of agricultural extension enable extension to operate in a systematic way. For extension to be truly effective, however, extension goals (and strategies to achieve these) must be continuously reevaluated and their implementation planned. Planning of extension activities takes place at different levels, but at each level there is the same concern for making general goals specific and for identifying strategies to achieve these goals. Agricultural extension and research staff dominate the extension planning process, but farmers and representatives of agricultural input and marketing agencies are involved at all levels. Effective planning depends on feedback from lower levels of strategy implementation; this is particularly important at planning levels closest to the farmer where objectives and strategies are the most detailed. Targets of activity are a vital part of the planning process, but care should be taken that the criteria for the targets are designed in terms of extension's specific goals.

19. *Agricultural Input Supply and Extension*

Agricultural input supply and agricultural extension are mutually dependent. Confusion over the responsibility of extension with respect to inputs is common although the relationship is clear: extension workers at any level do not handle any inputs and are not responsible for their distribution and sale. Extension does have an important role, however, in advising input agencies of the input supply situation in the field and anticipated demand. It also has an interest in the accuracy of this information as it will affect the timely availability of inputs. Farmers should not be advised of production recommendations involving inputs unless those inputs are available to them. To ensure this necessary coordination, representatives of input agencies participate in preseasonal, monthly, and fortnightly extension planning and training meetings.

20. *Training for Extension Staff*

Professional agricultural extension depends on the continuous upgrading of staff through training. As well as regular fortnightly, monthly, and seasonal training, specialized training of extension staff is required. Special training requirements should be determined under the guidance of extension management and Training Officers by evaluating the skills and training needs of all staff individually. The extension service must have a professional atmosphere in which a staff member who does not attend regular and specialized training and who does not improve professionally, feels out of place in the service's environment of learning, training, and know-how. Training priorities must be established, and long-term and annual programs to meet these drawn up. As a guide, each staff member should participate in at least one special short course each year. Some training requirements may be met through short courses held by universities and other training institutions. Others should be organized and given by extension staff. In addition to short courses, some staff will require longer-term upgrading through refresher training or university degree studies.

21. *Information Support*

Dissemination of information about extension operations is important within the extension service for the training and motivation of staff and as input for policy decisions. Externally, it builds up understanding and support for extension's objectives and achievements. Information support activities must be carefully planned, implemented, and monitored. Staff are available for the task, particularly Training Officers.

Success stories from staff, farmer reaction to extension, general data on extension's activities and achievements, and monitoring and evaluation results are chief subjects for dissemination. Suitable verified stories on these subjects should be publicized and made available to newspapers and radio and at exhibitions.

22. *Communication Techniques*

Without an effective system of communication within the extension service and between it and farmers, agricultural extension can achieve little. The training and visit system of extension establishes a broad structure to facilitate such communication, but equally important are the communication skills of extension staff. Five communication techniques have proved to be particularly effective in training extension staff: practical orientation, skill teaching, trainee involvement, samples and examples, and visual aids. These techniques have some application in extension's contact with farmers, but there the use of contact farmers and initial implementation of production recommendations on small areas are important communication techniques.

23. *Incentives for Extension Staff*

Staff of an agricultural extension service should receive appropriate incentives to work well. The most effective incentive to good work is that a job be purposeful and satisfying, and that good work be recognized and rewarded. Appropriate incentives are particularly important in extension, since the effectiveness of the system depends to a large extent on the contribution of lower-level staff. The employment structure of an extension service must provide material and intangible encouragement for staff at all levels. An attrative, flexible remuneration structure, access to training on the basis of need and ability, promotion in response to responsibility and ability, and the establishment of professional staff cadres, all under effective management, are necessary. Selective incentives such as awards and study tours are most useful in the context of a well-managed system that caters to the overall professional development of all extension staff.

24. *Agricultural Extension and Farm Women*

Women have an important role in agriculture. Their involvement in agriculture varies between cultures, but in most there are a few major agricultural operations in which women do not participate. To be truly effective, an agricultural extension service must deal with the activities handled by women. This is not often done, however, because of sociological constraints and inadequate focus by extension on women's agricultural activities. An agricultural extension service can adopt a number of strategies to improve its support for farm women, among others by developing with research suitable production recommendations for activities solely or largely performed by women, orienting staff to the activities and needs of farm women, having some Subject Matter Specialists (SMSs) concentrate on women's activities, and employing women in all positions in the extension service for which they are qualified. These steps are important, but by themselves achieve little unless extension staff regularly meet farm women. Ways to ensure this contact include the selection in each farmers' group of some female contact farmers. Whatever approach is adopted to have extension serve farm women, local agricultural, sociological, and administrative considerations should be taken into account, and involvement of extension, research, and training staff at all levels is required.

25. *The Training and Visit System and the Department of Agriculture*

When a training and visit (T&V) system of agricultural extension is initially established, the attention of extension management is rightly focused on the mechanics of the system. Sight should not be lost, however, of the need for flexibility in the extension system, and of the broader implications of the T&V approach for a Department of Agriculture and for agricultural development activities in general. The organization of an extension service must over time adapt to increasingly complex and sophisticated demands of farmers, technological and research developments, changes in the supply of agricultural inputs and market conditions, and to the changing ability of its staff. Adaptations can include changes in the operational organization of the extension system, increased upgrading and specialization of staff, the establishment of professional staff cadres, enhanced use of mass media to complement field extension work, and a broader range of operations covered by the extension service. The adoption of T&V extension has significant implications for a Department of Agriculture's functions, programs, and staff; farmer participation in development activities; agricultural support services; and relationships between farmers, extension, and research. Properly exploited by extension management, these implications can significantly broaden the impact of a professional extension system.

Annex: Work Responsibilities of Extension Staff

For any system of agricultural extension to be effective, it must be well organized and have a cohesive structure. All participants must perform, and be held accountable for, a basic set of duties. Because of variations in local administrative conditions, a common set of comprehensive job descriptions may not apply to all extension services. However, basic duties for each level of participants in the training and visit system can be defined. Whatever local adjustments are made, the duties of staff at each position should conform to those of the equivalent officer as elaborated here.

- —Director of Extension
- —Zonal Extension Officer (ZEO)
- —District Extension Officer (DEO)
- —Subject Matter Specialist (Plant Protection, Agronomy, Special Subjects, etc.) (SMS)
- —Training Officer
- —District (or University) Extension Agronomist (DEA)
- —Subdivisional Extension Officer (SDEO)
- —Agricultural Extension Officer (AEO)
- —Village Extension Worker (VEW)